Ethan Russo, MD
Editor

D0141061

Cannabis:
From Pariah
to Prescription

Cannabis: From Pariah to Prescription has been co-published simultaneously as *Journal of Cannabis Therapeutics*, Volume 3, Numbers 3 and 4 2003.

Pre-publication
REVIEWS,
COMMENTARIES,
EVALUATIONS . . .

"**E**DUCATIONAL AND PROVOCATIVE. . . . EXTREMELY TIMELY. . . . A compendium of clinical studies conducted by GW Pharmaceuticals using their cannabis-based medical extracts (CBME). Dr. Russo's introduction reviews the potential therapeutic benefits of cannabis for a variety of maladies. This chapter will be particularly interesting to readers wishing to understand the possible reasons for the differences between herbal cannabis and pharmaceutically prepared delta-9-THC. It also presents the fascinating history of the development of CBME."

Ken Mackie, MD
Professor
Anesthesiology
Adjunct Professor
Physiology & Biophysics
University of Washington
School of Medicine
Seattle

Cannabis:
From Pariah
to Prescription

Cannabis: From Pariah to Prescription has been co-published simultaneously as *Journal of Cannabis Therapeutics*, Volume 3, Numbers 3 and 4 2003.

The *Journal of Cannabis Therapeutics* Monographic "Separates"

Below is a list of "separates," which in serials librarianship means a special issue simultaneously published as a special journal issue or double-issue *and* as a "separate" hardbound monograph. (This is a format which we also call a "DocuSerial.")

"Separates" are published because specialized libraries or professionals may wish to purchase a specific thematic issue by itself in a format which can be separately cataloged and shelved, as opposed to purchasing the journal on an on-going basis. Faculty members may also more easily consider a "separate" for classroom adoption.

"Separates" are carefully classified separately with the major book jobbers so that the journal tie-in can be noted on new book order slips to avoid duplicate purchasing.

You may wish to visit Haworth's website at . . .

http://www.HaworthPress.com

. . . to search our online catalog for complete tables of contents of these separates and related publications.

You may also call 1-800-HAWORTH (outside US/Canada: 607-722-5857), or Fax 1-800-895-0582 (outside US/Canada: 607-771-0012), or e-mail at:

docdelivery@haworthpress.com

Cannabis: From Pariah to Prescription, edited by Ethan Russo, MD (Vol. 3, No. 3, and Vol. 3, No. 4, 2003). *This book reviews the latest research from recent clinical trials with cannabis and cannabinoids, outlining their place and future as prescription medicines.*

Women and Cannabis: Medicine, Science, and Sociology, edited by Ethan Russo, MD, Melanie Dreher, PhD, and Mary Lynn Mathre, RN, MSN (Vol. 2, No. 3/4, 2002). *Examines the therapeutic role of medicinal marijuana in women's medicine and its implications for fertility and maternal/child health.*

Cannabis Therapeutics in HIV/AIDS, edited by Ethan Russo, MD (Vol. 1, No. 3/4, 2001). *Explores delivery methods, clinical studies, and the history of cannabis therapy with HIV/AIDS patients.*

Cannabis:
From Pariah
to Prescription

Ethan Russo, MD
Editor

Cannabis: From Pariah to Prescription has been co-published simultaneously as *Journal of Cannabis Therapeutics*, Volume 3, Numbers 3 and 4 2003.

The Haworth Integrative Healing Press
The Haworth Herbal Press
Imprints of
The Haworth Press, Inc.

New York • London • Victoria (AU)
www.HaworthPress.com

LIBRARY
FRANKLIN PIERCE COLLEGE
RINDGE, NH 03461

Published by

The Haworth Integrative Healing Press®, 10 Alice Street, Binghamton, NY 13904-1580 USA

The Haworth Integrative Healing Press® is an imprint of The Haworth Press, Inc., 10 Alice Street, Binghamton, NY 13904-1580 USA.

Cannabis: From Pariah to Prescription has been co-published simultaneously as *Journal of Cannabis Therapeutics*, Volume 3, Numbers 3 and 4 2003.

© 2003 by The Haworth Press, Inc. All rights reserved. No part of this work may be reproduced or utilized in any form or by any means, electronic or mechanical, including photocopying, microfilm and recording, or by any information storage and retrieval system, without permission in writing from the publisher. Printed in the United States of America.

The development, preparation, and publication of this work has been undertaken with great care. However, the publisher, employees, editors, and agents of The Haworth Press and all imprints of The Haworth Press, Inc., including The Haworth Medical Press® and Pharmaceutical Products Press®, are not responsible for any errors contained herein or for consequences that may ensue from use of materials or information contained in this work. Opinions expressed by the author(s) are not necessarily those of The Haworth Press, Inc. With regard to case studies, identities and circumstances of individuals discussed herein have been changed to protect confidentiality. Any resemblance to actual persons, living or dead, is entirely coincidental.

Cover design by Brooke Stiles

Library of Congress Cataloging-in-Publication Data

Cannabis : from pariah to prescription / Ethan Russo.
     p. ; cm.
     "Co-published simultaneously as Journal of cannabis therapeutics, volume 3, numbers 3 and 4, 2003."
     Includes bibliographical references and index.
     ISBN 0-7890-2398-9 (hard cover : alk. paper)–ISBN 0-7890-2399-7 (soft cover : alk. paper)
     1. Cannabinoids–Therapeutic use. 2. Cannabis–Social aspects.
     [DNLM: 1. Cannabinoids–pharmacology. 2. Cannabinoids–therapeutic use. 3. Phytotherapy. 4. Placebos–administration & dosage. 5. Plant Extracts–therapeutic use. QV 77.7 C22478 2003] I. Russo, Ethan.
   RM666.C266C365 2003
   615'.7827–dc22
                                              2003021326

Indexing, Abstracting & Website/Internet Coverage

This section provides you with a list of major indexing & abstracting services. That is to say, each service began covering this periodical during the year noted in the right column. Most Websites which are listed below have indicated that they will either post, disseminate, compile, archive, cite or alert their own Website users with research-based content from this work. (This list is as current as the copyright date of this publication.)

Abstracting, Website/Indexing Coverage . . . . Year When Coverage Began

- *Analgesia File, Dannemiller Memorial Educational Foundation, Texas <http://www.pain.com>* 2001

- *Chemical Abstracts Services <http://www.cas.org>* 2001

- *CNPIEC Reference Guide: Chinese National Directory of Foreign Periodicals* 2001

- *CINAHL (Cumulative Index to Nursing & Allied Health Literature), in print, EBSCO, and SilverPlatter, Data-Star, and PaperChase. (Support materials include Subject Heading List, Database Search Guide, and instructional video). <http://www.cinahl.com>* 2001

- *Drug Policy Information Clearinghouse* 2001

- *e-psyche, LLC <http://www.e-psyche.net>* 2001

- *EMBASE/Excerpta Medica Secondary Publishing Division. Included in newsletters, review journals, major reference works, magazines & abstract journals. <http://www.elsevier.nl>* . 2002

- *Excerpta Medica . . . See EMBASE/Excerpta Medica* 2002

- *FACT, Focus on Alternative & Complementary Therapies <http://www.ex.ac.uk/FACT>* 2001

- *Greenfiles* 2001

- *Herb Research Foundation <http://www.herbs.org>* 2000

(continued)

Special Bibliographic Notes related to special journal issues (separates) and indexing/abstracting:

- indexing/abstracting services in this list will also cover material in any "separate" that is co-published simultaneously with Haworth's special thematic journal issue or DocuSerial. Indexing/abstracting usually covers material at the article/chapter level.
- monographic co-editions are intended for either non-subscribers or libraries which intend to purchase a second copy for their circulating collections.
- monographic co-editions are reported to all jobbers/wholesalers/approval plans. The source journal is listed as the "series" to assist the prevention of duplicate purchasing in the same manner utilized for books-in-series.
- to facilitate user/access services all indexing/abstracting services are encouraged to utilize the co-indexing entry note indicated at the bottom of the first page of each article/chapter/contribution.
- this is intended to assist a library user of any reference tool (whether print, electronic, online, or CD-ROM) to locate the monographic version if the library has purchased this version but not a subscription to the source journal.
- individual articles/chapters in any Haworth publication are also available through the Haworth Document Delivery Service (HDDS).

ABOUT THE EDITOR

Ethan Russo, MD, is a board-certified child and adult neurologist formerly with Montana Neurobehavioral Specialists in Missoula, MT. He is a researcher in migraine, ethnobotany, medicinal plants, cannabis and cannabinoids in pain management, and the therapeutic applications of Schedule I plants and chemicals. He serves as Senior Medical Advisor to GW Pharmaceuticals Group.

Dr. Russo holds faculty appointments as adjunct associate professor in the Department of Pharmaceutical Sciences of the University of Montana, and clinical associate professor in the Department of Medicine of the University of Washington.

He has published numerous articles in scientific journals and is the author of *Handbook of Psychotropic Herbs: A Scientific Analysis of Herbal Preparations for Psychiatric Conditions*. He is co-editor with Franjo Grotenhermen of the book, *Cannabis and Cannabinoids: Pharmacology, Toxicology and Therapeutic Potential*, and author of the novel, *The Last Sorcerer: Echoes of the Rainforest*, all from Haworth Press.

Dr. Russo is the founding editor of *Journal of Cannabis Therapeutics: Studies in Endogenous, Herbal & Synthetic Cannabinoids*, whose charter issue was released in January 2001. Two double-issues are also published as books, *Cannabis Therapeutics in HIV/AIDS*, and *Women and Cannabis: Medicine, Science, and Sociology*. He has published over two dozen articles on topics of neurology, clinical cannabis, and medicinal plants.

Dr. Russo has served as a consultant for private pharmaceutical companies, medical-legal cases, and in conservation policies with regards to medicinal herbs.

He lives in Missoula, MT surrounded by nature, is married to a pediatric nurse practitioner, and has two teenage children.

ALL HAWORTH INTEGRATIVE HEALING PRESS
BOOKS AND JOURNALS ARE PRINTED ON
CERTIFIED ACID-FREE PAPER

Cannabis:
From Pariah to Prescription

CONTENTS

Introduction:
Cannabis:
From Pariah to Prescription

Ethan Russo

SUMMARY. Cannabis has been employed in human medicine for more than 4000 years. In the last century, political prohibition led to its disappearance from the conventional pharmacopoeia, but this trend is reversing due to the broad acceptance and application of this forbidden medicine by patients with chronic and intractable disorders inadequately treated by available therapeutics. This study addresses the "road back" for cannabis medicines, and reacceptance as prescription products.

Current pharmacology of the two primary therapeutic phytocannabinoids, THC and CBD, is reviewed with respect to herbal synergy and as pertains to treatment of pain, spasm and the wide range of therapeutic applications and adverse effects of cannabis.

In particular, the efforts of GW Pharmaceuticals to develop cannabis based medicine extracts (CBME) are documented including cultivation of genetically-selected medical-grade cannabis cloned strains in glass houses with organic and integrated pest management techniques, and their processing employing supercritical carbon dioxide extraction and winterization. These CBMEs are then available for formulation of dos-

Ethan Russo, MD, is a Clinical Child and Adult Neurologist, Clinical Assistant Professor of Medicine, University of Washington, and Adjunct Associate Professor of Pharmacy, University of Montana, 2235 Wylie Avenue, Missoula, MT 59802 USA (E-mail: erusso@montanadsl.net).

[Haworth co-indexing entry note]: "Introduction: Cannabis: From Pariah to Prescription." Russo, Ethan. Co-published simultaneously in *Journal of Cannabis Therapeutics* (The Haworth Integrative Healing Press, an imprint of The Haworth Press, Inc.) Vol. 3, No. 3, 2003, pp. 1-29; and: *Cannabis: From Pariah to Prescription* (ed: Ethan Russo) The Haworth Integrative Healing Press, an imprint of The Haworth Press, Inc., 2003, pp. 1-29. Single or multiple copies of this article are available for a fee from The Haworth Document Delivery Service [1-800-HAWORTH, 9:00 a.m. - 5:00 p.m. (EST). E-mail address: docdelivery@haworthpress.com].

http://www.haworthpress.com/store/product.asp?sku=J175
© 2003 by The Haworth Press, Inc. All rights reserved.
10.1300/J175v03n03_01

1

age forms including sublingual extracts and inhaled forms. An optional Advanced Delivery System (ADS) is also discussed. *[Article copies available for a fee from The Haworth Document Delivery Service: 1-800-HAWORTH. E-mail address: <docdelivery@haworthpress.com> Website: <http://www. HaworthPress.com> © 2003 by The Haworth Press, Inc. All rights reserved.]*

KEYWORDS. Medical marijuana, cannabis, THC, cannabidiol, herbal treatment, alternative delivery systems, psychopharmacology

The *Journal of Cannabis Therapeutics* is pleased to mark with this the publication the transition of cannabis from a forbidden herb back into the realm of prescription medicine. Although a recognized and documented therapeutic agent for more than 4000 years (Aldrich 1997; Russo 2003; Russo 2001), cannabis became politicized in the early 20th century, leading to its ultimate prohibition in most industrialized nations. Cannabis was dropped from the *National Formulary* in the USA in 1941, and the *British Pharmacopoeia* in 1971. Reasons for the loss of cannabis as an available pharmaceutical were complex, related to a perceived risk of abuse, but also included formidable quality control issues such as lack of reliable or consistent supplies from India, idiosyncratic variability of patient responses to available preparations, and the advent of modern single product pharmacotherapy. The road back for cannabis medicines, as it were, has been a difficult and circuitous journey, beset by politics to a greater extent than science.

The essential features that characterize a prescription medicine require it to be of proven quality, consistency, clinical efficacy, and safety. For the last thirty-plus years, 85% of the world's research dollars for cannabis have been provided by the National Institute on Drug Abuse (NIDA), whose orientation has certainly not tended towards proof of therapeutic efficacy for this ancient herb. The lead has thus been taken by Europeans, whose medicine has never strayed quite so far from the realm of vegetable *materia medica*. Our account will document the progress of GW Pharmaceuticals, which, with full backing of the UK Home Office, has achieved the feat in five years of progressing from the idea of restoring cannabis to the pharmacy, all the way through to submission of a lead product for regulatory approval by the Medicines and Healthcare Products Regulatory Agency (MHRA, formerly the Medicines Control Agency).

As previously published two years ago (Whittle, Guy, and Robson 2001), many hurdles exist when considering the concept of how to produce a prescription cannabis product (p. 186):

- the concept of cannabis-based medicines as botanicals as opposed to pure cannabinoids;
- selective breeding of high yielding chemovars that produce an abundance of one particular cannabinoid;
- investigation of the pharmacological properties of various cannabinoids, i.e., cannabis is not just THC;
- variability of composition of cannabis. The geographical and genetic basis for variation in cannabinoid content of cannabis biomass and its control to give a standardised product;
- the quality aspects of cannabis biomass production;
- routes of administration and optimisation of formulations to achieve particular pharmacokinetic profiles;
- regulatory issues, including health registration, and international legal requirements;
- security packaging and anti-diversionary devices which can be used in connection with cannabis-based medicines in order to satisfy statutory requirements.

As is evident, the process of preparing a botanical for approval as medicine is comparable, but yet more complex than that for the New Chemical Entity (NCE), or novel synthetic pharmaceutical. A formidable barrier remains in the assignation and perception of cannabis as a drug of abuse. In the USA, cannabis was placed in the most restrictive category, Schedule I of the Controlled Substances Act in 1970, which encompasses drugs that are dangerous and addictive and lack recognized medical utility. It requires emphasis that this assignment was political and designed as a temporary, pending reassignment by the Shaffer Commission in 1972 (Abuse 1972). President Nixon rejected their recommendations of medical access and decriminalization before even reading the final report. Additionally, Schedule I assignation remains anachronistic (Haines et al. 2000). Many such drugs, including cannabis and LSD have had clear therapeutic indications in the past. Others, such as diamorphine (heroin), are forbidden in the USA, but retain legal pharmaceutical status in the UK. At least, controversy about such blanket proscriptions exists, and certainly with advancing knowledge, debate and reconsideration are required. A detailed analysis of the complexities of the cannabis question in the UK is available (Whittle and Guy

2003). The same publication outlines scientific evidence that cannabis based medicine extracts (CBME) may offer a distinct advantage over THC alone (Marinol®):

1. *Potentiation.* Based on a concept noted for endocannabinoids and their precursors called the "entourage effect" (Ben-Shabat et al. 1998; Mechoulam and Ben-Shabat 1999), various phytocannabinoid components, whether active (CBD, CBC) or relatively inactive (CBN) affect the cannabinoid receptor binding, pharmacokinetics and metabolism of THC. The same may be true of non-cannabinoid components, such as the essential oil terpenoids (McPartland and Russo 2001; Russo and McPartland 2003).
2. *Antagonism.* Cannabidiol mitigates side effects of THC (Karniol et al. 1975; Mechoulam, Parker, and Gallily 2002), including its intoxication liability. Additionally, other cannabis components may be helpful in this regard, e.g., terpenoids such as pulegone, 1, 8-cineole, and α-pinene may counter the short-term memory impairment engendered by THC (McPartland and Russo 2001; Russo and McPartland 2003).
3. *Summation.* A number of cannabis components may contribute to a certain therapeutic effect of THC (Williamson and Evans 2000; McPartland and Russo 2001).
4. *Pharmacokinetic.* For example, CBD alters the metabolism of THC by inhibiting its hepatic conversion to 11-OH-THC (Zuardi et al. 1982).
5. *Metabolism.* Whittle and Guy (2003) argue, as have others (Tyler 1994; Russo 2001) that due to co-evolution over the millennia, humans are better able to metabolize herbal preparations (i.e., cannabis) as compared to synthetic pharmaceuticals (i.e., synthetic cannabinoids).

Beyond the issues of regulation and rationale, the next step is to grow the plant. *Cannabis sativa*, despite its cosmopolitan propagation on the planet, is a rather exacting species insofar as optimal production of desirable medicinal cannabinoids is concerned. Such production is greatest in unfertilized female flowering tops, most commonly known as *sinsemilla* (Spanish, "without seed"), or *ganja*, the Sanskrit term for a process known in India for some 2500 years (Figure 1). THC production is increased by selecting certain strains and exposing them to ultraviolet light (Pate 1994). In the organization of the primary GW Pharmaceuticals production glasshouse, David Potter and Etienne de

FIGURE 1. Unfertilized female cannabis flower (photograph courtesy of GW Pharmaceuticals).

Reprinted with permission from GW Pharmaceuticals.

Meijer have outlined additional important factors (Potter 2003; de Meijer 2003): high yield per area, high cannabinoid purity, high inflorescence to leaf ratio ("harvest index"), avoidance of diseases and pests, production of sturdy growth conducive to subsequent processing and ease of harvest.

Consistency is achieved by clonal propagation of cuttings from select strains called "mother plants," that yield shorter specimens with less waste stem material. Successful propagation occurs with 95% of cuttings (Figure 2).

A decision was made to produce different cannabinoid ratios for prescription CBMEs, through the use of separate high-THC and high-CBD strains, or their combination in a fixed-ratio. This work was initiated by HortaPharm B.V. a generation ago in Holland, and selected strains were developed there, and the seeds imported into the UK in 1998 (de Meijer 2003). The high-THC strain was originally produced by hybridization of ((Afghani × Mexican) × Colombian) genetics, said to be reminiscent of the commercial (if illegal) "Skunk #1" strain (Potter 2003). An initial 400 plants grown from seed were analyzed for cannabinoid concentration and purity, leading to five chemovars ("chemical varieties" or phenotypes) that were selected for commercial cultivation potential. A high-CBD strain was similarly selected from 1600 seeds yielding a selection of the best four chemovars. It has been determined that cannabis

FIGURE 2. Clonal growth in glasshouse (photograph courtesy of GW Pharmaceuticals).

Reprinted with permission from GW Pharmaceuticals.

plant vigor, architecture, and glandular trichome density and metabolic efficiency in cannabinoid production are all polygenetically-determined traits, but affected by environmental factors (de Meijer 2003; de Meijer et al. 2003). Together, they determine the "cannabinoid quality." The chemovar is the primary determinant, however, of what cannabinoid ratios result. Additional line selection via repetitive self-fertilization has also been employed to maximize appropriate selection of both parents of a hybrid (de Meijer 2003).

In this particular instance, GW Pharmaceuticals chose to produce separate chemovars that selectively yield THC, CBD and THCV (83% theoretical maximum), CBC (76% theoretical maximum) or even CBG in relatively high amounts (de Meijer 2003). Although genetic modification (GM) of cannabis has often been discussed in certain quarters, it is abundantly clear from the above discussion that tremendous variation of chemical parameters is readily available with application of standard Mendelian genetic breeding techniques, and there is no rational reason for adding to the cannabis controversy by rendering it a genetically-modified organism (GMO).

As cannabis propagation and quality are subject to the vagaries of weather, all the more in a cloudy and wet northern clime, artificial lighting under glass was deemed the preferred method for pharmaceutical production in the UK. Mother plants are grown under high-pressure sodium (HPS) lights continuously at 75 watts/m^2 PAR (Photosythetically Active Radiation) (equivalent to 31,000 lux of natural sunlight) at 25°C in an organic compost ("leaf mould") to a height of 2 m, allowing pruning and the production of as many as 80 more cuttings for propagation (Potter 2003). The mother plant may be utilized for two or more "flushes" over the next few months before its vigor diminishes.

Clones are placed in peat pots after treatment with rooting hormone, trimming to retain one axial bud, and are grown out in polythene tunnels under high humidity with 24 hour light for two weeks until "potting up" (Potter 2003). Plants are continued under perpetual illumination for about three weeks until attaining a height of 50 cm, before shifting to a 12-hour light/12-hour dark critical day-length regimen to induce flowering.

All cultivation is performed in accord with Good Agricultural Practice (GAP) methods of the European Medicines Evaluation Agency in conjunction with rules of the UK Medicines Control Agency (Medicines Control 1997) for the production of a Botanical Drug Substance (BDS). [For the approved process of medicinal cannabis cultivation in the Netherlands, see the article in a prior issue of *Journal of Cannabis Therapeutics* (Anonymous 2003)]. Microbiological safety is crucial, and is a monitored function by regulatory agencies. In this instance GW Pharmaceuticals chose to use some minimal mineral sources of soil enrichment to avoid possible pathogen exposure from organic sources (Potter 2003). However, no pesticides whatsoever have been employed. Common pests are kept at bay by positive pressure in the glasshouses, and utilization of integrated pest management (IPM). Pests of concern have included spider mites (*Tetanychus* spp.) and onion (tobacco) thrips (*Thrips tabaci*). These are controlled through release of predatory mites, and kept at low level. For a comprehensive examination of the topic, the reader is urged to consult the superb *Hemp diseases and pests: Management and biological control* (McPartland, Clarke, and Watson 2000).

Fungal issues to date at the GW facilities have mainly pertained to grey mold (*Botrytis cinerea*) and powdery mildew (*Sphaerotheca macularis*). Control is achieved mainly by avoidance of high humidity close to time of harvest for the former, and increasing light pressure while avoiding excessive nitrogen exposure for the latter. When diseased plants do arise, affected specimens are destroyed.

While trials of outdoor cultivation were attempted with CBD-rich strains, daunting problems were encountered in the cool, damp British climate (Potter 2003).

Because cannabigerol (CBG) levels are dependent upon plant maturity, both the THC- and CBD-rich chemovars are harvested at the same growth stage at the onset of senescence, at which time the flowering tops representing 90% of the weight of the plants' aerial portions. Drying under a stream of dehumidified air from 25 down to 12% moisture content is then achieved under dark conditions to minimize cannabinoid oxidation (Whittle, Guy, and Robson 2001). The resultant mixture of dried unfertilized flowers, stalks and leaves yields 15% THC or 8% CBD in the respective chemovars (Figures 3, 4, and 5).

Interestingly, in the "raw" state, much of the THC and CBD are in the form of cannabinoid acids, THCA and CBDA, which are low in cannabinoid pharmacological activity. It is only after decarboxylation by progressive oxidation over time, after heating, or in the extraction process, that significant THC and CBD levels are produced and pharmacological benefits are obtained.

FIGURE 3. High CBD strain in GW Pharmaceuticals glasshouse (photograph courtesy of David Downs, PhD, GW Pharmaceuticals).

Reprinted with permission from GW Pharmaceuticals.

FIGURE 4. High THC strain in GW Pharmaceuticals glasshouse (photograph courtesy of David Downs, PhD, GW Pharmaceuticals).

Reprinted with permission from GW Pharmaceuticals.

Historically, cannabis extracts were ethanol-based, dating back to the experiments of O'Shaughnessy in India in the 19th century (O'Shaughnessy 1838-1840). GW Pharmaceuticals has opted for a more modern technique employing supercritical CO_2 extraction (Whittle, Guy, and Robson 2001). This has distinct advantages, as organic materials are extracted at approximately body temperature with retention of essential oil terpenoid components that seemingly contribute to medicinal effects of cannabis (McPartland and Russo 2001). Additionally, no solvent residue remains after the process. Although such extraction does include some waxy ballast, this is easily removed by "winterization," or chilling in an ethanol solution. The resultant liquid CBME is then ready for pharmaceutical preparation.

Whereas oral ingestion and smoking have been favored methods of application in the past, they are not likely to be the primary modes of administration in the future of clinical cannabis as a prescription medicine. Oral administration, such as with Marinol® (synthetic THC, or "dronabinol" in sesame oil) was introduced into the USA market in

FIGURE 5. Dried cannabis ready for processing (photograph courtesy of GW Pharmaceuticals).

Reprinted with permission from GW Pharmaceuticals.

1986, but has been relatively little employed (Russo 2002). Reasons include expense, delayed onset of effects in the range of 90-120 minutes, lack of practical titration of dosage, and a pronounced tendency toward dysphoria or other mental complaints from being "too high." In part, this may relate to hepatic first-pass conversion of THC to 11-OH-THC, which may possess a higher degree of psychoactivity according to some authorities. Interestingly, the presence of CBD, which is present in natural cannabis, but obviously absent in Marinol®, impedes this hepatic conversion by inhibition of cytochrome P450 3A11 (Browne and Weissman 1981).

Although smoking of cannabis was an acknowledged delivery system in the past, with cigarettes from the Grimault et Cie Company among others, and endorsement by such experts as Walter E. Dixon in England (Dixon 1899, 1921) and Walther Straub in Germany (Straub

1931), it is highly unlikely that regulatory agencies such as the Food and Drug Administration (FDA) would ever approve a drug delivery system that produces bronchial irritation and contains pyrolytic end-products that are potentially carcinogenic (Tashkin et al. 2002). Vaporization technology presents a viable option, preserving as it does the rapid bronchial absorption of cannabis components, and retaining the ability to titrate dosage rapidly. It is in initial stages of investigation (Gieringer 1996; Gieringer 1996; Gieringer 2001; Russo and Stortz 2003). This approach will require both elucidation of the pharmacokinetics of the vaporization technique, and approval of the hardware as a medical device. As will be discussed, GWP has developed an inhaled device (patent application GB0126150.2) that employs a metallic or ceramic surface coated with CBME that is heated by electrical current. The process is triggered by inhalation, employing the Advanced Dispensing System (ADS) (*vide infra*) providing the advantages of smoked cannabis (rapid onset, ready dosage titratability), but without hazards posed by smoke particles or inhalation of solvents.

Inhaled, non-smoked delivery of isolated THC has been previously investigated (Tashkin et al. 1977), but curiously, the isolated molecule is quite irritating to the bronchioles and induces a cough reflex despite its notable bronchodilatory benefits (Williams, Hartley, and Graham 1976). Biophysical parameters for this method of delivery are exacting, and have been recently reviewed (Whittle, Guy, and Robson 2001). Particles of diameter greater than 10 μ fail to reach the bronchioles. Those below 1 μ are mostly re-expired. It is only those particles in the 1-2 μ range that stand the best chance to be absorbed from the alveoli. Inasmuch as THC is an extremely viscous molecule that sticks to any vessel, dispersion in a solvent such as alcohol or propylene glycol is most often necessary, and introduces its own adverse effect issues in pulmonary application. This search for modern alternatives to smoked cannabis continues, however, through the use of a metered dose inhaler (Wilson et al. 2002) for THC. Although some seemingly represent that THC represents the sum total of important pharmacological effects of cannabis (Wachtel et al. 2002), others counter (McPartland and Russo 2001; Russo and McPartland 2003) in contrast, that the presence of other phytocannabinoid and terpenoid component such as myrcene, with its analgesic and anti-inflammatory effects (Rao, Menezes, and Viana 1990; Lorenzetti et al. 1991), or α-pinene, which is also a bronchodilator (Falk et al. 1990), or apigenin, wich is a non-sedating flavonoid in cannabis (Viola et al. 1995), contribute demonstrably to its clinical attributes. This debate will continue, engendering as it does the basic conflict between single-component "modern" pharmacology, and old-fashioned but resurgent notions of phytotherapeutic synergy.

Suppository forms of cannabis have been documented as far back as Ancient Egypt (Mannische 1989; Russo 2002), and the Victorian era (Farlow 1889). Modern research effort has also revived the concept, most often with Δ^9-THC-hemisuccinate (Broom et al. 2001; Elsohly et al. 1991). This method lacks convenience, is less subject to allow titration of dosage, and may be cosmetically unacceptable, especially in particular American consumers.

Transdermal delivery of cannabinoids is an attractive possibility as consumers have found "patches" to be a convenient method of drug delivery via this parenteral, long-acting method. Problems with this method have been previously outlined (Whittle, Guy, and Robson 2001). In essence, they include the lipophilic nature of cannabis components, the need for carrier molecules or other facilitators of transdermal absorption, and results to date that approximate only 10% of necessary serum levels (Challapalli and Stinchcomb 2002). Finally, the gradient of transport of cannabinoids through the skin is such that a used patch would still retain 90% or more of initial dosage, and would thereby represent a theoretical diversion risk upon disposal.

GW Pharmaceuticals primary efforts to date have focused on an approach employing a sub-lingual or oro-mucosal spray of CBME in ethanol and propylene glycol solution. The oro-mucosal preparation employs a pump action aerosol spray (Robson and Guy 2003; Whittle and Guy 2003) (Figure 6). This dispersion of materials allows reasonably rapid absorption (45 minutes), preserving the ability to titrate dosage, avoiding excessive swallowing of material, and producing an area under the curve that is comparable to that for smoked or intravenous administration of THC (Whittle and Guy 2003). Experiments in the UK with a simple unadorned device have demonstrated no major compliance problems, nor diversion of CBME to the black market. There are no plans to introduce pharmaceutical products with CBME in the UK, Western European or British Commonwealth nations with added security devices. However, it is anticipated that such security would be a necessary prerequisite in the USA for Drug Enforcement Administration (DEA) and Food and Drug Administration (FDA) approval (Figure 7). Thus, an additional Advanced Delivery System (ADS) has been developed (Figure 8). The ADS is a hand-held computerized encrypted device which may (Robson and Guy 2003; Whittle and Guy 2003):

1. remind patients of times dosing is due
2. record daily patterns and fluctuations in doses employed
3. allow remote computer monitoring of dosage employed by researchers or clinicians
4. render the device secure, tamper-proof, and patient-specific through individual codes
5. allow delivery of a variety of dosage forms (e.g., CBME with THC-CBD 1:1 ratio for daily usage, with high-THC preparation for sudden bouts of pain)
6. be suitable for usage with controlled drugs such as methadone or diamorphine (heroin).

CLINICAL STUDY DESIGN

As will be seen subsequently, initial Phase I studies of CBME examined pharmacokinetics and adverse effects of the materials in normal

FIGURE 6. Pump Action Sublingual Spray as utilized in the United Kingdom (photograph courtesy of GW Pharmaceuticals).

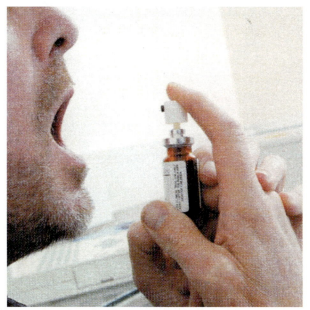

Reprinted with permission from GW Pharmaceuticals.

FIGURE 7. Sublingual spray as part of Advanced Delivery System (photograph courtesy of GW Pharmaceuticals).

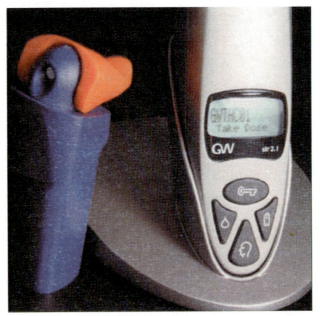

Reprinted with permission from GW Pharmaceuticals.

volunteers with monitoring of dose-response parameters, as well as pulse, blood pressure and subjective and objective assessments of intoxication. Although criticized, the current "gold standard" in pharmaceutical assessment is the double-blind randomized placebo-controlled clinical trial (RCT). An accepted variation in this approach that is worthwhile in contexts in which true blinding is difficult to achieve (as with cannabis) or in assessment of unpredictable diseases (such as multiple sclerosis) is presented by the N-of-1 trial design, achieved through a series of randomized, placebo controlled studies in which each subject serves as their own control (Guyatt et al. 1990). In fact, this approach to cannabis clinical trials was specifically endorsed by the American Institute of Medicine (Joy, Watson, and Benson 1999).

In assessing target conditions for initial studies, GWP relied on a survey of clinical cannabis patients and their conditions. In 1998, some 3516 self-selected patients who contacted the company were sent survey forms, of which 2458 were completed (70% response rate) (Robson and Guy 2003). Of 787 current or past cannabis users, the greatest rep-

FIGURE 8. Diagram of Advanced Delivery System (ADS) (courtesy of GW Pharmaceuticals).

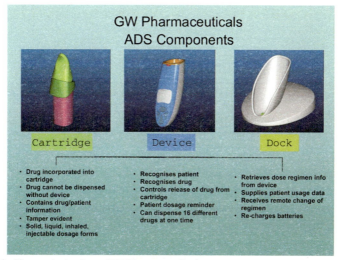

Reprinted with permission from GW Pharmaceuticals.

resentation was among patients with MS or various arthritic conditions. This contrasts with the situation in the USA, where HIV/AIDS is more highly represented, but where chronic pain remains a prime concern (Corral 2001; Gieringer 2001).

Another priority in selection of patients for clinical investigation involved a decision to study those with intractable conditions that had failed to be symptomatically controlled by available conventional pharmaceuticals. This was based on a philosophical decision to demonstrate that CBME would not merely be equal in efficacy to standard drugs, but rather, offer tangible advantages in difficult clinical contexts. A decision was also made to add CBME to patients' existing pharmaceutical regimens to provide a baseline comparison.

For MS patients, entry criteria included the presence of one or more poorly controlled symptom despite best available treatment: pain, spasm, spasticity, tremor or urinary difficulty, whether frequency, urgency, nocturia or incontinence (Robson and Guy 2003; Whittle and Guy 2003). Patient exclusions were similar to those employed in previous studies of cannabis or Marinol®: history of serious drug or alcohol abuse, schizophrenia, uncontrolled cardiovascular conditions including hypertension, impaired hepatic or renal function, and epilepsy. All national norms for clinical research and Guidelines for Good Clinical Practice (GCP) were followed.

Noting past historical data on individual idiosyncrasies of dosing and responses to cannabis by patients, a requirement was pursued to deliver initial THC:CBD 1:1 CBME dosages in open-label fashion with close monitoring, in an attempt to establish initial individual dose guidelines. This was then followed by randomized double-blind crossover comparisons of that preparation versus placebo and high-THC and high-CBD CBMEs. Subsequent monitoring employed an array of subjective measures (via visual analogue scales, or VAS) and objective measures on examination and laboratory study. Patients who demonstrated benefits in initial studies were given the option of entering long-term safety studies, and a majority of patient-subjects chose to do so (Robson and Guy 2003). The results of these trials form the basis for the remainder of this publication.

THC AND CBD: SUMMARY OF CURRENT KNOWLEDGE

Although this author has emphasized the biochemical and physiological contribution importance of other cannabis components (minor cannabinoids, terpenoids and flavonoids) to the medical therapeutic benefits of cannabis (McPartland and Russo 2001), it is clear from the data that exist to date that two entities provide the greatest effects: Δ^9-tetrahydrocannabinol and cannabidiol. A complete analysis of current knowledge is beyond our scope, but it is appropriate to briefly summarize current knowledge of their contributions (Table 1).

Receptor Effects

THC is a partial agonist at both CB_1 and CB_2 receptors (Pertwee 1998; Showalter et al. 1996). In contrast, CBD has little activity, and perhaps slight antagonistic activity at CB_1, while greater activity at CB_2 (Showalter et al. 1996). Of great importance, it has recently been demonstrated that cannabidiol stimulates vanilloid receptors (VR_1) with similar efficacy to capsaicin, and inhibits uptake of the endocannabinoid anandamide (AEA), and weakly inhibits its hydrolysis (Bisogno et al. 2001). These new findings have important implications in elucidating the pain-relieving and anti-inflammatory effects of CBD. In a

TABLE 1. Therapeutic/Adverse Effects of THC and CBD

Effect	THC	CBD	Reference
Receptor/Non-Receptor Effects			
CB_1 (CNS receptors)	++	±	(Pertwee 1998)
CB_2 (Peripheral receptors)	+	++	(Showalter et al. 1996)
Vanilloid Receptors	–	+	(Bisogno et al. 2001)
Anti-inflammatory	+	+	(Hampson et al. 1998)
Immunomodulatory	+	++	(Malfait et al. 2000; Cabral 2001)
CNS Effects			
Anticonvulsant	+	++	(Wallace, Martin, and DeLorenzo 2002; Carlini and Cunha 1981)
Muscle Relaxant	+	++	(Petro 1980)
Antinociceptive	++	+	(Pertwee 2001)
Catalepsy	++	++	(O'Shaughnessy 1838-1840)
Psychotropic	++	–	(Russo 2001)
Anxiolytic		+	(Zuardi and Guimaraes 1997)
Antipsychotic	–	++	(Zuardi and Guimaraes 1997)
Neuroprotective antioxidant activity	+	++	(Hampson et al. 1998)
Antiemetic	++	–	(Chang et al. 1979)/(Guy et al. 2002)
Sedation (reduced spontaneous activity)	+	+	(Zuardi and Guimaraes 1997)
Agitation (Alzheimer disease)	+		(Volicer et al. 1997)
Tic reduction	+		(Müller-Vahl et al. 1999)
Withdrawal effects (reduction)	+		(Cichewicz and Welch 2002; Reynolds 1890)
Migraine	+		(Russo 2001; Russo 1998)
Bipolar disease	+		(Grinspoon and Bakalar 1998)
Cardiovascular Effects			
Bradycardia	–	+	(Weil, Zinberg, and Nelsen 1968)
Tachycardia	+	–	ditto
Hypertension	+	–	ditto
Hypotension	–	+	(Adams et al. 1977)
Appetite/Gastrointestinal			
Appetite	+	–	(da Orta 1913)
Motility (slowed)	+		(Pertwee 2001)
Neonatal feeding (endocannabinoid)	+		(Fride 2002)
Anti-Carcinogenesis			
Melanoma (apoptosis, angiogenesis)	+		(Casanova et al. 2003)
Breast (prolactin receptor)	+		(De Petrocellis et al. 1998)
Glioma (apoptosis)	+	+	(Sanchez et al. 1998; Vaccani, Massi, and Parolaro 2003)
Leukemia (apoptosis)	+		(McKallip et al. 2002)
Pulmonary (blocks carcinogenesis enzymatically)	+		(Roth et al. 2001)
Pulmonary			
Bronchodilation	+		(Williams, Hartley, and Graham 1976; Tashkin et al. 1977)
Ophthalmological			
Intra-ocular pressure (reduced)	++	+	(Merritt et al. 1980; Jarvinen, Pate, and Laine 2002)
Night vision (improved)	+*		(Russo et al. 2003; West 1991)

Adapted and expanded from (Whittle, Guy, and Robson 2001; Whittle and Guy 2003).
* New indication. See final article in this publication.

manner of interpretation, CBD may be considered the first clinical agent that modulates endocannabinoid function.

Anti-Inflammatory and Immunomodulatory Effects

The benefits of cannabis and cannabinoids on inflammation have been extensively documented. The following are suggested as reviews (Hampson et al. 1998; Pertwee 2001; Burstein 1992; Russo 2001). Both THC and CBD have important roles in these observations. Of increasing interest is the recent demonstration that CBD possesses both anti-inflammatory and immunomodulatory benefits in an animal model of rheumatoid arthritis (Malfait et al. 2000). Although there has been great concern expressed as to immunological damage by cannabis, such effects are usually demonstrable in laboratory assays at levels 50-100 times the psychoactive dose (Cabral 2001). Deleterious clinical effects of cannabis in HIV (Abrams et al. 2002), and chronic medical usage (Russo et al. 2002) have not been demonstrated.

Central Nervous System Effects

Of prime importance in cannabinoid therapeutics is pain control or antinociception (Pertwee 2001; Russo 2001). One of the primary functions of the endogenous cannabinoid system is modulation of pain control, in parallel with the endogenous opioid and vanilloid systems. THC is the main contributor of cannabis to control of pain, via its actions on CB_1, which occur in key areas of the spinal cord, and brainstem. A purported "comprehensive" review of the analgesic effects of cannabinoids concluded that they have little demonstrated benefit (Campbell et al. 2001), but this pronouncement produced strong refutation (Russo 2001) and more considered subsequent support (Baker et al. 2003) in some quarters. Countless testimonials attest to the unique benefits of cannabis in difficult cases of neuropathic pain (Grinspoon and Bakalar 1997), and other unusual and intractable conditions, such as familial Mediterranean fever (Holdcroft et al. 1997).

The cataleptic effects of high doses of THC were noted by O'Shaughnessy in 1839 (O'Shaughnessy 1838-1840), and this effect remains part of the tetrad of behavioral effects sought in laboratory animals as a sign of cannabinoid activity.

Cannabis was noted to have anticonvulsant effects in the 19th century. Primary focus of therapeutic benefit on seizures of partial onset has focused on CBD (Carlini and Cunha 1981), while it was generally believed that THC was proconvulsant. Epileptic patients have generally claimed otherwise (Corral 2001), and it was recently demonstrated that endocannabinoids modulate seizure thresholds, and that THC exerts an anticonvulsant effect, as well (Wallace, Martin, and DeLorenzo 2002).

Migraine is a neurochemical and vascular disorder of exceeding complexity, whose treatment remains extremely problematical. The multi-modality effects of cannabis seem to support its historical role in both symptomatic and prophylactic treatment (Russo 1998; Russo 2001). While THC has received the bulk of the attention in therapeutic application, this author's experience with Marinol® treatment would seem to support that the benefits on chronic migraine treatment do not mirror the high efficacy of historical claims in the Victorian era. Current discoveries of the endocannabinoid modulation and vanilloid receptor effects of CBD discussed above (Bisogno et al. 2001) would seem to support that cannabidiol is a necessary component to successful prophylaxis in migraine.

Antidepressant and anti-anxiety effects of cannabis date to ancient India in the *Atharva Veda*, and the Scythians (Herodotus 1998). Certainly, an antidepressant effect of cannabis has been observed in chronic disease (Herodotus 1998; Russo et al. 2002; Regelson et al. 1976). In general, THC is considered psychotropic, while CBD generally is not (reviewed in Russo 2001). Rather, cannabidiol is noteworthy for its anxiolytic, sedative and antipsychotic effects (Zuardi and Guimaraes 1997). Interestingly, THC (as Marinol®) was recently observed to produce weight gain and reduce agitation in demented Alzheimer disease patients (Volicer et al. 1997). Unfortunately, CBD was not examined, but very likely would have contributed to the clinical benefits. Anecdotal reports support benefit of THC in mood-stabilization in bipolar disease (Grinspoon and Bakalar 1998).

The antispasmodic effects of cannabis were observed in such diseases as tetanus in the 19th century, producing cures of fatal diseases, and palliation of chronic disorders (O'Shaughnessy 1838-1840). Muscle relaxant properties of cannabis in multiple sclerosis were noted more recently (Petro 1980; Grinspoon and Bakalar 1997), and have recently been reviewed in detail (Baker et al. 2003; Consroe 1998; Petro 2002). These will form the focus of many of the study results subsequently discussed in this publication. As if the muscle relaxant and anti-spasmodic benefits of cannabis were insufficient, it has recently been demonstrated that cannabinoid agonists positively influence the immunological parameters of demyelinating diseases such as experimentally allergic encephalomyelitis (Baker et al. 2000). In the past year, a small clinical trial of THC and a cannabis extract was performed with 16 subjects. Neither was observed to reduce spasticity, and adverse events were reported in the extract group (Killestein et al. 2002). Numerous criticisms were subsequently voiced in this regard (Russo 2003). Among these were that the plant extract was poorly categorized; in fact, it contained a fixed ratio of THC to CBD with maximum doses of 5 mg of THC and 2 mg of CBD per day. The study additionally employed oral administration with no real dose titration. An additional study in Switzerland with more patients (57) and doses of up to 15 mg THC with 6 mg CBD divided tid has provided better results with reduction in spasms to the $p < 0.05$ level and no significant side effects vs. placebo (Vaney et al. 2002). A study of an even larger cohort of MS patients in the UK is pending publication.

Kirsten Müller-Vahl has pioneered the use of cannabis and THC in Tourette syndrome, demonstrating a marked reduction in tic behavior and obsessive-compulsive preoccupation (Muller-Vahl et al. 2003; Müller-Vahl et al. 1999).

The antiemetic effect of THC in morning sickness was noted as early as the 19th century (Wright 1862), and was further elucidated in the last two decades (Chang et al. 1979). A tremendous body of knowledge in this context that has been historically ignored was recently published in this journal (Musty and Rossi 2001). This pertained to state-sponsored studies in the USA in cancer chemotherapy. Pooling available data in some 768 patients, oral THC provided 76-88% relief of nausea and vomiting, while smoked cannabis figures supported 70-100% relief in the various surveys. Also worthy of inclusion here, an Israeli study of 8 children receiving highly emetogenic chemotherapy for hematological malignancies with oral Δ^8-THC (a trace and more stable component of cannabis) was 100% effective in allaying vomiting in 480 dose applications! Surprisingly, slight euphoria was noted in only one subject, causing the authors to surmise that the appreciation of the cannabis "high" is a developmental phenomenon. Shockingly, this study has never been followed by more similar investigations.

Surprisingly as well, it has just been demonstrated that CBD also has anti-emetic benefits in motion sickness in rodents (Guy et al. 2002), an indication that has wide implications, including space flight.

Although THC and cannabis are often attacked as productive of addiction, it is well documented from the 19th century that prominent physicians claimed benefit of Indian hemp in treatment of alcohol, morphine and cocaine dependencies (Reynolds 1890). As is becoming a recurrent theme, the claims of the Victorian era are resonating with modern scientists who subsequently prove their biochemical and physiological basis. This benefit has been strikingly demonstrated in the laboratory, through "opiate-sparing" by THC (Cichewicz et al. 1999), and more recently, the effect of THC to mitigate opiate-withdrawal symptoms, and block the formation of dependency (Cichewicz and Welch 2002).

One of the most exciting and pressing areas of neurological investigation surrounds the emerging concept of neuroprotection. If one were able to prevent the progressive cell death of parkinsonism, amyotrophic lateral sclerosis, Alzheimer and Huntington diseases, the inevitable deterioration and ultimate demise that these disorders eventuate might well be mitigated or arrested. This is the promise that may accrue to THC and CBD from the research of Hampson et al. (1998) in their demonstration that these agents are capable of blocking NMDA receptors in glutamate toxicity.

Cardiovascular Effects

A pioneering study in 1968 documented transient tachycardia and hypertension induced by THC in experimental subjects (Weil, Zinberg, and Nelsen 1968). Overall however, a mild hypotensive effect of CBD is observed (Adams et al. 1977). Recently, concerns have been raised with respect to cannabis as an inciting influence in myocardial infarction (Mittleman et al. 2001), but no significant epidemiological basis is evident for such claims (Sidney et al. 1997).

Appetite/Gastrointestinal

The appetite stimulating power of cannabis and THC are among the most well known effects (or side effects). This phenomenon was first documented in the West by the physician and explorer, Garcia da Orta, in India in the 16th century (da Orta 1913), but repeatedly studied subsequently. It was this effect that led to an approved indication for THC (as Marinol®) in the USA in 1992. Recently, smoked cannabis and THC demonstrated benefits in appetite and weight gain in hospitalized AIDS subjects (Abrams et al. 2002).

THC slows gut motility (reviewed in Pertwee 2001), providing additional support to the known analgesic and anti-inflammatory benefits in such disorders as Crohn's disease, ulcerative colitis, and idiopathic bowel syndrome (spastic colon).

A much better understanding of the critical role of tonic endocannabinoid function in normal ontogeny has recently been elucidated when Ester Fride and colleagues investigated the role of anandamide in initiation of neonatal feeding, and inevitable demise with its blockade (Fride 2002). Therapeutic use in "failure-to-thrive" states and cystic fibrosis (Fride 2002) are obvious putative applications.

Anti-Carcinogenesis

Whereas, governmental pronouncements have long sought to indict marijuana and THC as contributors to the incidence of cancer, closer analysis has failed to demonstrate epidemiological support for significant danger, even with smoked cannabis (Ware and Tawfik 2001). Little publicity, in contrast, has accrued to an increasing number of studies that demonstrate anti-carcinogenesis by THC.

Legitimate concerns surround the use of smoked cannabis, and its contribution to pulmonary irritation, bronchitis symptoms, and possible neoplastic sequelae (Tashkin 2001). However, recent study indicates that THC and even cannabis smoke block the activity of a key enzyme in pulmonary carcinogenesis (Roth et al. 2001), perhaps explaining the observation that there are still no documented cases of lung cancer in cannabis-only smokers.

THC also has been demonstrated to promote apoptosis (programmed cell death) in malignant conditions including: leukemia (McKallip et al. 2002) via CB_2 stimulation, gliomas (Sanchez et al. 1998), and melanoma (Casanova et al. 2003), in which tumor angiogenesis is also inhibited. Additionally, two types of breast tumor cell lines were inhibited by THC (De Petrocellis et al. 1998), apparently via prolactin receptor effects. This is obviously a fertile area for further research.

Pulmonary

As noted above, the primary medical concerns about cannabis revolve around its pulmonary sequelae. It requires emphasis that these may be totally avoided through alternative delivery techniques. That notwithstanding, it seems that emphysematous deterioration, even in cannabis smokers, is a lower risk than previously surmised (Tashkin et al. 1997). Actual therapeutic application of THC in asthma, as previously attempted (Tashkin et al. 1977; Williams, Hartley, and Graham 1976), may soon become a reality with improved vaporizers or CBME applications.

Ophthalmological

The ability of cannabis and THC to lower intra-ocular pressure in glaucoma was serendipitously discovered in the late 1970s by a variety of patients and researchers (Randall and O'Leary 1998; Merritt et al. 1980). What is more compelling perhaps, in the long run, is the fact that there is more to glaucoma treatment than merely controlling pressure. Even effective management with conventional pharmacology fails to avert visual loss over time. Rather, an emerging concept supports that prospect that glaucoma represents a progressive vascular retinopathy that requires a neuroprotectant to preserve vision (Jarvinen, Pate, and Laine 2002). This is an area where cannabis and cannabinoids shine.

As will be discussed in the final entry in this publication, cannabis and cannabinoids also seem to have a role in improving night vision and in treatment of other degenerative eye conditions (Russo et al. 2003).

REFERENCES

Abrams, D., R. Leiser, J. Hilton, S. Shade, T. Elbeik, F. Aweeka, N. Benowitz, B. Bredt, B. Kosel, J. Aberg, S. Deeks, T. Mitchell, K. Mulligan, J. McCune, and M. Schambelan. 2002. Short-term effects of cannabinoids in patients with HIV-1 infection. Paper read at Symposium on the Cannabinoids, July 13, at Asilomar Conference Center, Pacific Grove, CA.

Abuse, United States Commission on Marihuana and Drug. 1972. *Marihuana: A signal of misunderstanding; First report.* Washington: U.S. Govt. Print. Off.

Adams, M. D., J. T. Earnhardt, B. R. Martin, L. S. Harris, W. L. Dewey, and R. K. Razdan. 1977. A cannabinoid with cardiovascular activity but no overt behavioral effects. *Experientia* 33 (9):1204-5.

Aldrich, M. R. 1997. History of therapeutic cannabis. In *Cannabis in medical practice: A legal, historical and pharmacological overview of the therapeutic use of marijuana,* edited by M. L. Mathre. NC: McFarland.

Anonymous. 2003. Guidelines for cultivating cannabis for medicinal purposes [*Voorschriften voor de verbouw van cannabis voor medicinale doeleinden*]. *J Cannabis Therapeutics* 3 (2):51-61.

Baker, D., G. Pryce, J. L. Croxford, P. Brown, R. G. Pertwee, J. W. Huffman, and L. Layward. 2000. Cannabinoids control spasticity and tremor in a multiple sclerosis model. *Nature* 404 (6773):84-7.

Baker, D., G. Pryce, G. Giovannoni, and A. J. Thompson. 2003. The therapeutic potential of cannabis. *Lancet Neurology* 2 (May):291-298.

Ben-Shabat, S., E. Fride, T. Sheskin, T. Tamiri, M. H. Rhee, Z. Vogel, T. Bisogno, L. De Petrocellis, V. Di Marzo, and R. Mechoulam. 1998. An entourage effect: inactive endogenous fatty acid glycerol esters enhance 2-arachidonoyl-glycerol cannabinoid activity. *Eur J Pharmacol* 353 (1):23-31.

Bisogno, T., L. Hanus, L. De Petrocellis, S. Tchilibon, D. E. Ponde, I. Brandi, A. S. Moriello, J. B. Davis, R. Mechoulam, and V. Di Marzo. 2001. Molecular targets for cannabidiol and its synthetic analogues: effect on vanilloid VR1 receptors and on the cellular uptake and enzymatic hydrolysis of anandamide. *Br J Pharmacol* 134 (4):845-52.

Broom, S. L., K. J. Sufka, M. A. Elsohly, and R. A. Ross. 2001. Analgesic and reinforcing proerties of delta9-THC-hemisuccinate in adjuvant-arthritic rats. *J Cannabis Therapeutics* 1 (3-4):171-182.

Browne, R. G., and A. Weissman. 1981. Discriminative stimulus properties of delta 9-tetrahydrocannabinol: mechanistic studies. *J Clin Pharmacol* 21 (8-9 Suppl): 227S-234S.

Burstein, S. 1992. Eicosanoids as mediators of cannabinoid action. In *Marijuana/ Cannabinoids: Neurobiology and neurophysiology of drug abuse*, edited by L. Murphy and A. Bartke. Boca Raton: CRC Press.

Cabral, G. 2001. Immune system. In *Cannabis and cannabinoids: Pharmacology, toxicology and therapeutic potential*, edited by F. Grotenhermen and E. B. Russo. Binghamton, NY: Haworth Press.

Campbell, F. A., M. R. Tramber, D. Carroll, D. J. M. Reynolds, R. A. Moore, and H. J. McQuay. 2001. Are cannabinoids an effective and safe option in the management of pain? A qualitative systematic review. *Brit Medl J* 323 (7 July):1-6.

Carlini, E. A., and J. M. Cunha. 1981. Hypnotic and antiepileptic effects of cannabidiol. *J Clin Pharmacol* 21 (8-9 Suppl):417S-427S.

Casanova, M. L., C. Blazquez, J. Martinez-Palacio, C. Villanueva, M. J. Fernandez-Acenero, J. W. Huffman, J. L. Jorcano, and M. Guzman. 2003. Inhibition of skin tumor growth and angiogenesis in vivo by activation of cannabinoid receptors. *J Clin Invest* 111 (1):43-50.

Challapalli, P. V., and A. L. Stinchcomb. 2002. In vitro experiment optimization for measuring tetrahydrocannabinol skin permeation. *Int J Pharm* 241 (2):329-39.

Chang, A. E., D. J. Shiling, R. C. Stillman, N. H. Goldberg, C. A. Seipp, I. Barofsky, R. M. Simon, and S. A. Rosenberg. 1979. Delta-9-tetrahydrocannabinol as an antiemetic in cancer patients receiving high-dose methotrexate. A prospective, randomized evaluation. *Ann Intern Med* 91 (6):819-24.

Cichewicz, D. L., Z. L. Martin, F. L. Smith, and S. P. Welch. 1999. Enhancement of mu opioid antinociception by oral delta9-tetrahydrocannabinol: Dose-response analysis and receptor identification. *J Pharmacol Exp Ther* 289 (2):859-67.

Cichewicz, D. L., and S. P. Welch. 2002. The effects of oral administration of delta-9-THC on morphine tolerance and physical dependence. Paper read at Symposium on the Cannabinoids, July 13, at Asilomar Conference Center, Pacific Grove, CA.

Consroe, P. 1998. Brain cannabinoid systems as targets for the therapy of neurological disorders [In Process Citation]. *Neurobiol Dis* 5 (6 Pt B):534-51.

Corral, V.L. 2001. Differential effects of medical marijuana based on strain and route of administration: A three-year observational study. *J Cannabis Therapeutics* 1 (3-4):43-59.

da Orta, Garcia. 1913. *Colloquies on the simples and drugs of India*. London: Henry Sotheran.

de Meijer, E. 2003. The breeding of cannabis cultivars for pharmaceutical end uses. In *Medicinal uses of cannabis and cannabinoids*, edited by B. A. Whittle, G. W. Guy and P. Robson. London: Pharmaceutical Press.

de Meijer, E. P., M. Bagatta, A. Carboni, P. Crucitti, V. M. Moliterni, P. Ranalli, and G. Mandolino. 2003. The inheritance of chemical phenotype in *Cannabis sativa* L. *Genetics* 163 (1):335-46.

De Petrocellis, L., D. Melck, A. Palmisano, T. Bisogno, C. Laezza, M. Bifulco, and V. Di Marzo. 1998. The endogenous cannabinoid anandamide inhibits human breast cancer cell proliferation. *Proc Natl Acad Sci U S A* 95 (14):8375-80.

Dixon, W. E. 1899. The pharmacology of *Cannabis indica*. *Brit Med J* 2:1354-1357.

_____1921. A manual of pharmacology. 5th ed. London: Edward Arnold & Co.

Elsohly, M. A., T. L. Little, Jr., A. Hikal, E. Harland, D. F. Stanford, and L. Walker. 1991. Rectal bioavailability of delta-9-tetrahydrocannabinol from various esters. *Pharmacol Biochem Behav* 40 (3):497-502.

Falk, A. A., M. T. Hagberg, A. E. Lof, E. M. Wigaeus-Hjelm, and Z. P. Wang. 1990. Uptake, distribution and elimination of alpha-pinene in man after exposure by inhalation. *Scand J Work Environ Health* 16 (5):372-8.

Farlow, J. W. 1889. On the use of belladonna and *Cannabis indica* by the rectum in gynecological practice. *Boston Med Surg Jl* 120:507-509.

Fride, E. 2002. Cannabinoids and cystic fibrosis: A novel approach. *J Cannabis Therapeutics* 2 (1):59-71.

_____2002. Cannabinoids and feeding: The role of the endogenous cannabinoid system as a trigger for newborn suckling. *J Cannabis Therapeutics* 2 (3-4):51-62.

Gieringer, D. 1996. Why marijuana smoke harm reduction? *Bull Multidisciplinary Assoc Psychedelic Stud* 6 (64-66).

Gieringer, D. 1996. Waterpipe study. *Bull Multidisciplinary Assoc Psychedelic Stud* 6:59-63.

_____2001. Medical use of cannabis: Experience in California. In *Cannabis and cannabinoids: Pharmacology, toxicology, and therapeutic potential*, edited by F. Grotenhermen and E. Russo. Binghamton, NY: Haworth Press.

Gieringer, D. H. 2001. Cannabis "vaporization": A promising strategy for smoke harm reduction. *J Cannabis Therapeutics* 1 (3-4):153-170.

Grinspoon, L., and J. B. Bakalar. 1998. The use of cannabis as a mood stabilizer in bipolar disorder: anecdotal evidence and the need for clinical research. *J Psychoactive Drugs* 30 (2):171-7.

Grinspoon, L., and J. B. Bakalar. 1997. *Marihuana, the forbidden medicine*. Rev. and exp. ed. New Haven: Yale University Press.

Guy, G. W., B. A. Whittle, F. A. Javid, C. Wright, and R. J. Naylor. 2002. An inhibitory role for cannabinoids in the control of motion sickness in *Suncus marinus*. Paper read at Symposium on the Cannabinoids, at Asilomar Conference Center, Pacific Grove, CA.

Guyatt, G. H., J. L. Keller, R. Jaeschke, D. Rosenbloom, J. D. Adachi, and M. T. Newhouse. 1990. The n-of-1 randomized controlled trial: Clinical usefulness. Our three-year experience. *Ann Intern Med* 112 (4):293-9.

Haines, T., C. D. Adler, T. P. Farley, E. B. Russo, L. Grinspoon, and R. W. Sweet. 2000. Living with our drug policy. *Fordham Urban Law Jl* 28 (1):92-129.

Hampson, A. J., M. Grimaldi, J. Axelrod, and D. Wink. 1998. Cannabidiol and (-)delta9-tetrahydrocannabinol are neuroprotective antioxidants. *Proc Natl Acad Sci U S A* 95 (14):8268-73.

Herodotus. 1998. *The histories.* Translated by R. Waterfield and C. Dewald. Oxford [England]; New York: Oxford University Press.

Holdcroft, A., M. Smith, A. Jacklin, H. Hodgson, B. Smith, M. Newton, and F. Evans. 1997. Pain relief with oral cannabinoids in familial Mediterranean fever. *Anaesthesia* 52 (5):483-6.

Jarvinen, T., D. Pate, and K. Laine. 2002. Cannabinoids in the treatment of glaucoma. *Pharmacol Ther* 95 (2):203.

Joy, Janet E., Stanley J. Watson, and John A. Benson, Jr. 1999. Marijuana and medicine: Assessing the science base. Washington, DC: Institute of Medicine.

Karniol, I. G., I. Shirakawa, R. N. Takahashi, E. Knobel, and R. E. Musty. 1975. Effects of delta9-tetrahydrocannabinol and cannabinol in man. *Pharmacol* 13 (6):502-12.

Killestein, J., E. L. Hoogervorst, M. Reif, N. F. Kalkers, A. C. Van Loenen, P. G. Staats, R. W. Gorter, B. M. Uitdehaag, and C. H. Polman. 2002. Safety, tolerability, and efficacy of orally administered cannabinoids in MS. *Neurol* 58 (9):1404-7.

Lorenzetti, B. B., G. E. Souza, S. J. Sarti, D. Santos Filho, and S. H. Ferreira. 1991. Myrcene mimics the peripheral analgesic activity of lemongrass tea. *J Ethnopharmacol* 34 (1):43-8.

Malfait, A. M., R. Gallily, P. F. Sumariwalla, A. S. Malik, E. Andreakos, R. Mechoulam, and M. Feldmann. 2000. The nonpsychoactive cannabis constituent cannabidiol is an oral anti-arthritic therapeutic in murine collagen-induced arthritis. *Proc Natl Acad Sci U S A* 97 (17):9561-6.

Mannische, L. 1989. *An ancient Egyptian herbal.* Austin: University of Texas.

McKallip, R. J., C. Lombard, M. Fisher, B. R. Martin, S. Ryu, S. Grant, P. S. Nagarkatti, and M. Nagarkatti. 2002. Targeting CB2 cannabinoid receptors as a novel therapy to treat malignant lymphoblastic disease. *Blood* 100 (2):627-34.

McPartland, J. M., R. C. Clarke, and D. P. Watson. 2000. *Hemp diseases and pests: Management and biological control.* Wallingford, UK: CABI.

McPartland, J. M., and E. B. Russo. 2001. Cannabis and cannabis extracts: Greater than the sum of their parts? *J Cannabis Therapeutics* 1 (3-4):103-132.

Mechoulam, R., and S. Ben-Shabat. 1999. From gan-zi-gun-nu to anandamide and 2-arachidonoylglycerol: The ongoing story of cannabis. *Nat Prod Rep* 16 (2): 131-43.

Mechoulam, R., L. A. Parker, and R. Gallily. 2002. Cannabidiol: an overview of some pharmacological aspects. *J Clin Pharmacol* 42 (11 Suppl):11S-19S.

Medicines Control, Agency. 1997. Rules and guidance for pharmaceutical manufacturers 1997: Orange Guide. London: Stationery Office Agency.

Merritt, J. C., W. J. Crawford, P. C. Alexander, A. L. Anduze, and S. S. Gelbart. 1980. Effect of marihuana on intraocular and blood pressure in glaucoma. *Ophthalmol* 87 (3):222-8.

Mittleman, M. A., R. A. Lewis, M. Maclure, J. B. Sherwood, and J. E. Muller. 2001. Triggering myocardial infarction by marijuana. *Circulation* 103 (23):2805-9.

Müller-Vahl, K. R., U. Schneider, H. Kolbe, and H. M. Emrich. 1999. Treatment of Tourette's syndrome with delta-9-tetrahydrocannabinol. *Am J Psychiatry* 156 (3):495.

Muller-Vahl, K. R., U. Schneider, H. Prevedel, K. Theloe, H. Kolbe, T. Daldrup, and H. M. Emrich. 2003. Delta9-tetrahydrocannabinol (THC) is effective in the treatment of Tics in Tourette syndrome: A 6-week randomized trial. *J Clin Psychiatry* 64 (4):459-465.

Musty, R. E., and R. Rossi. 2001. Effects of smoked cannabis and oral delta-9-tetrahydrocannabinol on nausea and emesis after cancer chemotherapy: A review of state clinical trials. *J Cannabis Therapeutics* 1 (1):29-42.

O'Shaughnessy, W. B. 1838-1840. On the preparations of the Indian hemp, or gunjah *(Cannabis indica)*; Their effects on the animal system in health, and their utility in the treatment of tetanus and other convulsive diseases. *Transactions of the Medical and Physical Society of Bengal*:71-102, 421-461.

Pate, D. 1994. Chemical ecology of cannabis. *J Interntl Hemp Assoc* 2:32-37.

Pertwee, R. G. 2001. Cannabinoid receptors and pain. *Prog Neurobiol* 63 (5):569-611.

———2001. Cannabinoids and the gastrointestinal tract. *Gut* 48 (6):859-67.

Pertwee, R.G. 1998. Advanced in cannabinoid receptor pharmacology in cannabis. In *Cannabis: The genus Cannabis*, edited by D. T. Brown. Amsterdam: Harwood Academic Publishers.

Petro, D. J. 1980. Marihuana as a therapeutic agent for muscle spasm or spasticity. *Psychosomatics* 21 (1):81, 85.

Petro, D. J. 2002. Cannabis in multiple sclerosis: Women's health concerns. *J Cannabis Therapeutics* 2 (3-4):161-175.

Potter, D. 2003. Growth and morphology of medicinal cannabis. In *Medicinal uses of cannabis and cannabinoids*, edited by B. A. Whittle, G. W. Guy and P. Robson. London: Pharmaceutical Press.

Randall, R. C., and A. M. O'Leary. 1998. *Marijuana Rx: The patients' fight for medicinal pot*. New York: Thunder's Mouth Press.

Rao, V. S., A. M. Menezes, and G. S. Viana. 1990. Effect of myrcene on nociception in mice. *J Pharm Pharmacol* 42 (12):877-8.

Regelson, W., J. R. Butler, J. Schulz, T. Kirk, L. Peek, M. L. Green, and M. O. Zalis. 1976. Delta 9-tetrahydrocannabinol as an effective antidepressant and appetite-stimulating agent in advanced cancer patients. In Braude M. C., Szara S., ed. *Pharmacology of marihuana*. Vol 2. New York, Raven Press.

Reynolds, J. R. 1890. Therapeutical uses and toxic effects of *Cannabis indica*. *Lancet* 1:637-638.

Robson, P., and G. W. Guy. 2003. Clinical studies of cannabis-based medicine. In *Medicinal uses of cannabis and cannabinoids*, edited by B. A. Whittle, G. W. Guy and P. Robson. London: Pharmaceutical Press.

Roth, M. D., J. A. Marques-Magallanes, M. Yuan, W. Sun, D. P. Tashkin, and O. Hankinson. 2001. Induction and regulation of the carcinogen-metabolizing enzyme CYP1A1 by marijuana smoke and delta (9)-tetrahydrocannabinol. *Am J Respir Cell Mol Biol* 24 (3):339-44.

Russo, E. 1998. Cannabis for migraine treatment: The once and future prescription? An historical and scientific review. *Pain* 76 (1-2):3-8.

———2001. Cannabinoids in pain management. Study was bound to conclude that cannabinoids had limited efficacy. *Brit Med J* 323 (7323):1249-50.

_____2002. Cannabis treatments in obstetrics and gynecology: A historical review. *J Cannabis Therapeutics* 2 (3-4):5-35.

Russo, E. B., and J. M. McPartland. 2003. Cannabis is more than simply delta (9)-tetrahydrocannabinol. *Psychopharmacol (Berl)* 165 (4):431-2.

Russo, E. B. 2001. *Handbook of psychotropic herbs: A scientific analysis of herbal remedies for psychiatric conditions.* Binghamton, NY: Haworth Press.

_____2002. Role of cannabis and cannabinoids in pain management. In *Pain management: A practical guide for clinicians,* edited by R. S. Weiner. Boca Raton, FL: CRC Press.

_____2003. The history of cannabis as medicine. In *Medicinal uses of cannabis and cannabinoids,* edited by B. A. Whittle, G. W. Guy and P. Robson. London: Pharmaceutical Press.

Russo, E. B., M. L. Mathre, A. Byrne, R. Velin, P. J. Bach, J. Sanchez-Ramos, and K. A. Kirlin. 2002. Chronic cannabis use in the Compassionate Investigational New Drug Program: An examination of benefits and adverse effects of legal clinical cannabis. *J Cannabis Therapeutics* 2 (1):3-57.

Russo, E. B., A. Merzouki, J. Molero Mesa, and K. A. Frey. 2003. Cannabis improves night vision: A pilot study of dark adaptometry and scotopic sensitivity in kif smokers of the Rif Mountains of Northern Morocco. *Journal of Ethnopharmacology* (Submitted).

Russo, E.B., and M. Stortz. 2003. An interview with Markus Storz: June 19, 2002. *J Cannabis Therapeutics* 3 (1):67-78.

Russo, E. B. 2001. Hemp for headache: An in-depth historical and scientific review of cannabis in migraine treatment. *J Cannabis Therapeutics* 1 (2):21-92.

Sanchez, C., I. Galve-Roperh, C. Canova, P. Brachet, and M. Guzman. 1998. Delta9-tetrahydrocannabinol induces apoptosis in C6 glioma cells. *FEBS Lett* 436 (1):6-10.

Showalter, V. M., D. R. Compton, B. R. Martin, and M. E. Abood. 1996. Evaluation of binding in a transfected cell line expressing a peripheral cannabinoid receptor (CB2): Identification of cannabinoid receptor subtype selective ligands. *J Pharmacol Exp Ther* 278 (3):989-99.

Sidney, S., J. E. Beck, I. S. Tekawa, C. P. Quesenberry, and G. D. Friedman. 1997. Marijuana use and mortality. *Am J Public Health* 87 (4):585-90.

Straub, W. 1931. Intoxicating drugs. In *Lane Lectures on Pharmacology,* edited by W. Straub. Stanford, CA: Stanford University Press.

Tashkin, D. P., S. Reiss, B. J. Shapiro, B. Calvarese, J. L. Olsen, and J. W. Lodge. 1977. Bronchial effects of aerosolized delta 9-tetrahydrocannabinol in healthy and asthmatic subjects. *Am Rev Respir Dis* 115 (1):57-65.

Tashkin, D. P., M. S. Simmons, D. L. Sherrill, and A. H. Coulson. 1997. Heavy habitual marijuana smoking does not cause an accelerated decline in FEV1 with age. *Am J Respir Crit Care Med* 155 (1):141-8.

Tashkin, D. P., G. C. Baldwin, T. Sarafian, S. Dubinett, and M. D. Roth. 2002. Respiratory and immunologic consequences of marijuana smoking. *J Clin Pharmacol* 42 (11 Suppl):71S-81S.

Tashkin, D. P. 2001. Respiratory risks from marijuana smoking. In *Cannabis and cannabinoids: Pharmacology, toxicology and therapeutic potential,* edited by F. Grotenhermen and E. Russo. Binghamton, NY: Haworth Press.

Tyler, V. E. 1994. *Herbs of choice: the therapeutic use of phytomedicinals.* New York: Pharmaceutical Products Press.

Vaccani, A., P. Massi, and D. Parolaro. 2003. Inhibition of human glioma cell growth by the non psychoactive cannabidiol. Paper read at First European Workshop on Cannabinoid Research., April 4-5, at Madrid.

Vaney, C., P. Jobin, F. Tschopp, M. Heinzel, and M. Schnelle. 2002. Efficacy, safety and tolerability of an orally administered cannabis extract in the treatment of spasticity in patients with multiple sclerosis. Paper read at Symposium on the Cannabinoids, July 13, at Asilomar Conference Center, Pacific Grove, CA.

Viola, H., C. Wasowski, M. Levi de Stein, C. Wolfman, R. Silveira, F. Dajas, J. H. Medina, and A. C. Paladini. 1995. Apigenin, a component of *Matricaria recutita* flowers, is a central benzodiazepine receptors-ligand with anxiolytic effects. *Planta Med* 61 (3):213-6.

Volicer, L., M. Stelly, J. Morris, J. McLaughlin, and B. J. Volicer. 1997. Effects of dronabinol on anorexia and disturbed behavior in patients with Alzheimer's disease. *Int J Geriatr Psychiatry* 12 (9):913-9.

Wachtel, S.R., M. A. ElSohly, R. A. Ross, J. Ambre, and H. de Wit. 2002. Comparison of the subjective effects of delta9-tetrahydrocannabinol and marijuana in humans. *Psychopharmacol* 161:331-339.

Wallace, M. J., B. R. Martin, and R. J. DeLorenzo. 2002. Evidence for a physiological role of endocannabinoids in the modulation of seizure threshold and severity. *Eur J Pharmacol* 452 (3):295-301.

Ware, M. A., and V. L. Tawfik. 2001. A review of the respiratory effects of smoked cannabis: Implications for clinical trials. Paper read at Symposium on the Cannabinoids, June 30, at El Escorial, Spain.

Weil, A. T., N. E. Zinberg, and J. M. Nelsen. 1968. Clinical and psychological effects of marihuana in man. *Science* 162 (859):1234-42.

West, M.E. 1991. Cannabis and night vision. *Nature* 351:703-704.

Whittle, B. A., and G. W. Guy. 2003. Development of cannabis-based medicines; risk, benefit and serendipity. In *Medicinal uses of cannabis and cannabinoids*, edited by B. A. Whittle, G. W. Guy and P. Robson. London: Pharmaceutical Press.

Whittle, B. A., G. W. Guy, and P. Robson. 2001. Prospects for new cannabis-based prescription medicines. *J Cannabis Therapeutics* 1 (3-4):183-205.

Williams, S. J., J. P. Hartley, and J. D. Graham. 1976. Bronchodilator effect of delta1-tetrahydrocannabinol administered by aerosol of asthmatic patients. *Thorax* 31 (6):720-3.

Williamson, E. M., and F. J. Evans. 2000. Cannabinoids in clinical practice. *Drugs* 60 (6):1303-14.

Wilson, D. M., J. Peart, B. R. Martin, D. T. Bridgen, P. R. Byron, and A. H. Lichtman. 2002. Physiochemical and pharmacological characterization of a delta (9)-THC aerosol generated by a metered dose inhaler. *Drug Alcohol Depend* 67 (3):259-67.

Wright, T. L. 1862. Correspondence. *Cincinnati Lancet and Observer* 5 (4):246-247.

Zuardi, A. W., I. Shirakawa, E. Finkelfarb, and I. G. Karniol. 1982. Action of cannabidiol on the anxiety and other effects produced by delta 9-THC in normal subjects. *Psychopharmacol* 76 (3):245-50.

Zuardi, A. W., and F. S. Guimaraes. 1997. Cannabidiol as an anxiolytic and antipsychotic. In *Cannabis in medical practice: A legal, historical and pharmacological overview of the therapeutic use of marijuana*, edited by M. L. Mathre. Jefferson, NC: McFarland.

GW Pharmaceuticals List
of Abbreviations and Definitions of Terms

Abbreviation/Term	Definition/Explanation
°C	Degrees Celsius
AE	Adverse Event
$AUC_{0-\infty}$	The area under the plasma concentration versus time curve from zero to t calculated as AUC_{0-t} plus the extrapolated amount from time t to infinity
AUC_{0-t}	The area under the plasma concentration versus time curve, from time zero to 't' (where t = the final time of positive detection) as calculated by the linear trapezoidal method
ALT	Alanine Aminotransferase
ANOVA	Analysis of Variance
AST	Aspartate Aminotransferase
BMI	Body Mass Index; Weight (kg)/Height (m^2)
BP	Blood Pressure (systolic and diastolic)
BS-11	Box Scale 11
CBD	Cannabidiol
CBME	Cannabis Based Medicine Extract
CI	Confidence Interval

Reprinted with permission from GW Pharmaceuticals.

[Haworth co-indexing entry note]: "GW Pharmaceuticals List of Abbreviations and Definitions of Terms." Co-published simultaneously in *Journal of Cannabis Therapeutics* (The Haworth Integrative Healing Press, an imprint of The Haworth Press, Inc.) Vol. 3, No. 3, 2003, pp. 31-33; and: *Cannabis: From Pariah to Prescription* (ed: Ethan Russo) The Haworth Integrative Healing Press, an imprint of The Haworth Press, Inc., 2003, pp. 31-33. Single or multiple copies of this article are available for a fee from The Haworth Document Delivery Service [1-800-HAWORTH, 9:00 a.m. - 5:00 p.m. (EST). E-mail address: docdelivery@haworthpress.com].

http://www.haworthpress.com/store/product.asp?sku=J175
10.1300/J175v03n03_02

Abbreviation/Term	Definition/Explanation
C_{max}	Maximum measured plasma concentration
CPMP	Committee for Proprietary Medicinal Products
CRF	Case Report Form
CV%	Coefficient of Variation
DBP	Diastolic Blood Pressure
ECG	Electrocardiogram
Eth	Ethanol
GGT	Gamma Glutamyl Transferase
GCP	Good Clinical Practice
GW	GW Pharma Ltd/GW Pharmaceuticals Ltd
Hb	Haemoglobin
ICH	International Conference of Harmonisation
K_{el}	The Elimination Rate Constant
LLOQ	Lower Limit of Quantification
MCH	Mean Corpuscular Haemoglobin
MCHC	Mean Corpuscular Haemoglobin Concentration
MCV	Mean Cell Volume
MedDRA	Medical Dictionary for Regulatory Activities
N	Number
PG	Propylene Glycol
pg	Picogram
PK	Pharmacokinetic
ppmt	Peppermint
PR	PR segment in the tracing on the electrocardiogram
QT	QT segment in the tracing on the electrocardiogram
QT_c	QT segment in the tracing on the electrocardiogram corrected for breathing
QRS	QRS segment in the tracing on the electrocardiogram
RCC	Red Cell Count

Abbreviation/Term	Definition/Explanation
SBP	Systolic Blood Pressure
SD	Standard Deviation
SOC	System Organ Class
SOP	Standard Operating Procedure
THC	Δ^9-Tetrahydrocannabinol
T_{max}	Time to the Maximum Measured Plasma Concentration
$t_{1/2}$	Half Life
WCC	White Cell Count

A Single Centre, Placebo-Controlled, Four Period, Crossover, Tolerability Study Assessing, Pharmacodynamic Effects, Pharmacokinetic Characteristics and Cognitive Profiles of a Single Dose of Three Formulations of Cannabis Based Medicine Extracts (CBMEs) (GWPD9901), Plus a Two Period Tolerability Study Comparing Pharmacodynamic Effects and Pharmacokinetic Characteristics of a Single Dose of a Cannabis Based Medicine Extract Given via Two Administration Routes (GWPD9901 EXT)

G. W. Guy
M. E. Flint

G. W. Guy and M. E. Flint are affiliated with GW Pharmaceuticals plc, Porton Down Science Park, Salisbury, Wiltshire, SP4 0JQ, UK.

[Haworth co-indexing entry note]: "A Single Centre, Placebo-Controlled, Four Period, Crossover, Tolerability Study Assessing, Pharmacodynamic Effects, Pharmacokinetic Characteristics and Cognitive Profiles of a Single Dose of Three Formulations of Cannabis Based Medicine Extracts (CBMEs) (GWPD9901), Plus a Two Period Tolerability Study Comparing Pharmacodynamic Effects and Pharmacokinetic Characteristics of a Single Dose of a Cannabis Based Medicine Extract Given via Two Administration Routes (GWPD9901 EXT)." Guy, G. W., and M. E. Flint. Co-published simultaneously in *Journal of Cannabis Therapeutics* (The Haworth Integrative Healing Press, an imprint of The Haworth Press, Inc.) Vol. 3, No. 3, 2003, pp. 35-77; and: *Cannabis: From Pariah to Prescription* (ed: Ethan Russo) The Haworth Integrative Healing Press, an imprint of The Haworth Press, Inc., 2003, pp. 35-77. Single or multiple copies of this article are available for a fee from The Haworth Document Delivery Service [1-800-HAWORTH, 9:00 a.m. - 5:00 p.m. (EST). E-mail address: docdelivery@haworthpress.com].

http://www.haworthpress.com/store/product.asp?sku=J175
© 2003 by The Haworth Press, Inc. All rights reserved.
10.1300/J175v03n03_03

SUMMARY. This study was the first study of GW's CBME in man. It was performed in six healthy subjects, employing test treatments consisting of CBD:THC sublingual drops (GW-1011-01): 5 mg Δ^9-tetrahydrocannabinol (THC) + 5 mg cannabidiol (CBD) per ml of glycerol:ethanol (Eth):propylene glycol (PG) (4:4:2), with peppermint flavouring, High CBD sublingual drops (GW-3009-01): 5 mg CBD per ml of glycerol: Eth:PG (4:4:2), with peppermint flavouring, High THC sublingual drops (GW-2009-01): 5 mg THC per ml of glycerol:Eth:PG (4:4:2), with peppermint flavouring, placebo sublingual drops (GW-4003-01): glycerol: Eth:PG (4:4:2), with peppermint flavouring, aerosol (GW-1009-01): 5 mg CBD + 5 mg THC per ml formulated in propellant:Eth (80:20), and nebuliser (GW-1012-01): 10 mg CBD + 10 mg THC per ml of cremophor (Crem) (0.4):PG (1.5):macrogol (1):dodecanol (0.8):H_2O (7.4), and placebo nebuliser (administered to subjects 005 and 006 instead of the active nebuliser test treatment): Crem (0.4):PG (1.5):macrogol (1):dodecanol (0.8):H_2O (7.4).

Periods 1, 5 and 6 were open label, Periods 2 to 4 double blind. The study was a partially randomised crossover using single doses of THC and/or CBD or placebo. The study drug was administered as sublingual drops according to a pre-determined randomisation scheme in Periods 1 to 4. In Period 5, CBD:THC was administered as a sublingual aerosol and in Period 6 CBD:THC was administered as an inhalation via a nebuliser. There was a six-day washout between each dose.

Primary objectives of the study were to make a preliminary evaluation of the tolerability of cannabis based medicine extracts at single dose in comparison to placebo in order to provide guidance for dosage in future studies; GWPD9901 EXT: was designed to compare the effect of method of administration (sublingually via an aerosol) or the route (inhalation) on the cannabis based medicine extract containing THC and CBD in a ratio of 1:1 in terms of subjective assessment of well-being, *in vivo* pharmacokinetic characteristics over 12 h, the adverse event (AE) profile and measurement of vital signs and conjunctival reddening over 12 h.

Secondary objectives were to compare the effects of the four preparations in terms of cognitive assessment, subjective assessment of well-being *in vivo* pharmacokinetic characteristics over 12 h, the AE profile and measurement of vital signs and conjunctival reddening over 12 h.

The methodology was a six single dose, partially randomised, six-way cross-over study. In Period 1, all subjects received CBD:THC drops. In Periods 2-4, High THC drops, High CBD drops and placebo drops were administered, double blind and fully randomised. In Period 5, all subjects received the aerosol test treatment and in Period 6, all subjects received the nebuliser test treatment.

Each subject received five single doses of a maximum of 20 mg CBD, 20 mg CBD + 20 mg THC and 20 mg THC on five separate occasions and a placebo dose on one occasion. The duration of the study was six weeks.

Following administration of CBD:THC (Sativex) sublingual drops, mean concentrations of CBD, THC and 11-hydroxy-THC were above the Lower Limit of Quantification (LLOQ) by 45 min post-dose. Plasma concentrations of THC were at least double those of CBD before both decreased below the LLOQ by 360 min and 480 min post-dose, respectively. When High CBD sublingual drops were administered, plasma levels of CBD were generally similar to those measured after CBD:THC sublingual drops. High THC resulted in marginally earlier detection of mean concentrations of both THC and 11-hydroxy-THC and a slightly earlier decline than for CBD:THC sublingual plasma concentrations. Following administration of CBD:THC via the pressurised aerosol, mean quantifiable levels of CBD and THC were detected marginally earlier than for the CBD:THC sublingual drops and declined below the LLOQ marginally later. Plasma concentrations of THC, 11-hydroxy-THC and CBD following administration via the aerosol were lower than after administration of the sublingual drops. Following administration of CBD:THC via the nebuliser, mean plasma levels of both CBD and THC increased rapidly (within 5 min) to levels much higher than measured following administration of the sublingual drops and were maintained until around 120 min post-dose before declining rapidly. Levels of 11-hydroxy-THC were very low compared with those after sublingual dosing.

There were no statistically significant differences in the pharmacokinetics of THC or CBD between CBD:THC sublingual drops and High THC, High CBD or pressurised aerosol. With the exception of a single statistically significant difference in $AUC_{0-\infty}$ for 11-hydroxy-THC following administration of the High THC compared with CBD:THC sublingual drops there were no significant differences in the PK of 11-hydroxy-THC either.

Dosing with the inhaled nebuliser produced marked differences in the pharmacokinetics of CBD and THC compared with CBD:THC sublingual dosing. Peak concentration was greater and much earlier although only C_{max} of CBD and T_{max} of THC were statistically significantly different. Peak concentration and AUCs of 11-hydroxy-THC were statistically significantly less, reflecting reduced early metabolism of THC by this route.

No consistent statistically significant differences were noted between the pharmacokinetic parameters of High CBD, High THC and the aerosol when compared to the CBD:THC sublingual drops. However, the nebuliser resulted in a rapid absorption of CBD and THC and higher peak plasma levels but a reduction in the metabolism of THC to 11-hydroxy-THC.

Subjects experienced a reduction in wakefulness, feeling of well-being, mood, production of saliva and increased hunger and unpleasant effect following administration of each test treatment and placebo. The maximum mean changes in wakefulness, feeling of well-being, mood and production of saliva were reported 3 h post-dose following administration of CBD:THC sublingual drops. Similar trends were also reported following administration of placebo and therefore it is suggested that the effects reported may not be entirely due to active test treatments. The greatest mean incidence of unpleasant effects was reported earlier than for any other effect and following administration of the nebuliser test treatment.

The sublingual test treatments were best liked and the nebuliser test treatment was least liked. All of the subjects (100%) reported coughing and three subjects (50%) reported a sore throat following dosing with the nebuliser.

The sublingual test treatments were well tolerated by all subjects. All six subjects experienced at least two AEs during the study, but there were no deaths, serious adverse events (SAEs) or other significant AEs. The commonest AEs were tachycardia, conjunctival hyperaemia and abnormal dreams.

The small variations in individual subject laboratory parameters and urinalyses and in the mean laboratory parameters did not suggest any patterns or trends. The mean values of all the vital signs showed no patterns or trends either and no differences from placebo. ECGs at both screening and post-study were normal for all subjects.

In conclusion, each sublingual test treatment was well tolerated by all subjects. The inhaled test treatment was not well tolerated and resulted in adverse effects. *[Article copies available for a fee from The Haworth Document Delivery Service: 1-800-HAWORTH. E-mail address: <docdelivery@ haworthpress.com> Website: <http://www.HaworthPress.com> © 2003 by The Haworth Press, Inc. All rights reserved.]*

KEYWORDS. Cannabinoids, cannabis, THC, cannabidiol, medical marijuana, pharmacokinetics, pharmacodynamics, multiple sclerosis, botanical extracts, alternative delivery systems, harm reduction

INTRODUCTION

Cannabis plants (*Cannabis sativa*) contain approximately 60 different cannabinoids (British Medical Association 1997) and in the UK, oral tinctures of cannabis were prescribed until cannabis was made a

Schedule 1 controlled substance in the Misuse of Drugs Act in 1971. The prevalence of recreational cannabis use increased markedly in the UK after 1960, reaching a peak in the late 1970s. This resulted in a large number of individuals with a range of intractable medical disorders being exposed to the drug, and many of these discovered that cannabis could apparently relieve symptoms not alleviated by standard treatments. This was strikingly the case with certain neurological disorders, particularly multiple sclerosis (MS). The black market cannabis available to those patients is thought to have contained approximately equal amounts of the cannabinoids Δ^9-tetrahydrocannabinol (THC) and cannabidiol (CBD) (Baker, Gough, and Taylor 1983). The importance of CBD lies not only in its own inherent therapeutic profile but also in its ability to modulate some of the undesirable effects of THC through both pharmacokinetic and pharmacodynamic mechanisms (McPartland and Russo 2001). MS patients claimed beneficial effects from cannabis in many core symptoms, including pain, urinary disturbance, tremor, spasm and spasticity (British Medical Association 1997). The MS Society estimated in 1998 that up to 4% (3,400) of UK MS sufferers used cannabis medicinally (House of Lords 1998).

Cannabinoid clinical research has often focussed on synthetic analogues of THC, the principal psychoactive cannabinoid, given orally. This has not taken the possible therapeutic contribution of the other cannabinoid and non-cannabinoid plant components into account, or the slow and unpredictable absorption of cannabinoids via the gastrointestinal tract (Agurell et al. 1986). Under these conditions it has been difficult to titrate cannabinoids accurately to a therapeutic effect. Research involving plant-derived material has often reported only the THC content (Maykut 1985) of the preparations, making valid comparisons between studies difficult.

GW has developed cannabis based medicine extracts (CBMEs) derived from plant cultivars that produce high and reproducible yields of specified cannabinoids. CBMEs contain a defined amount of the specified cannabinoid(s), plus the minor cannabinoids and also terpenes and flavonoids. The specified cannabinoids constitute at least 90% of the total cannabinoid content of the extracts. The minor cannabinoids and other constituents add to the overall therapeutic profile of the CBMEs and may play a role in stabilising the extract (Whittle, Guy, and Robson 2001). Early clinical studies indicated that sublingual dosing with CBME was feasible, well tolerated and convenient for titration. The concept of self-titration was readily understood by patients and worked well in practice. Dosing patterns tended to resemble those seen in the patient controlled analgesia technique used in post-operative pain control; with small doses administered as and when patients require them, up to a maximal rate and daily limit (GW Pharmaceuticals 2002). The Phase 2 experience has supported some of the wide-range of effects reported anecdotally for cannabis. It has also shown that for most patients the therapeutic benefits of CBMEs could be obtained at doses below those that cause marked intoxication (the 'high'). This is consistent with experience in patients receiving opioids for pain relief, where therapeutic use rarely leads to misuse (Portenoy 1990; Porter and Jick 1980). Onset of intoxication may be an indicator of over-titration. However the range of daily dose required is subject to a high inter-individual variability.

The CBME GW-1000-02 is administered as an oromucosal spray, and contains an equal proportion of THC and CBD, similar to the cannabinoid profile of the cannabis thought to be most commonly available on the European black market (Baker, Gough, and Taylor 1983). The CBME GW-2000-02 is administered as an oromucosal spray, and contains over 90% THC. In this study, the CBME was administered sublingually as drops (GW-1011-01, GW-3009-01, GW-2009-01 and GW-4003-01), a pressurised aerosol (GW-1009-01) and as an inhalation via a nebuliser (GW-1012-01). Each formulation contained either equal amounts of CBD and THC, CBD alone or THC alone.

GWPD9901 was a Phase I clinical study designed to investigate the tolerability, cognitive effects, pharmacokinetic (PK) and pharmacodynamic (PD) effects of CBD and THC when co-administered and administered alone. It was also designed to assess safety and tolerability of the test treatments. It was the first exposure in man of GW's CBME formulations.

STUDY OBJECTIVES

Primary objectives of GWPD9901 were to make a preliminary evaluation of the tolerability of cannabis based medicine extracts (CBMEs) at single dose in comparison to placebo in order to provide guidance for dosage in future studies; while in GWPD9901 EXT: they were to compare the effect of method of administration (sublingually via an aerosol) or the route (inhalation) on the cannabis based medicine extract containing THC and CBD in a ratio of 1:1 in terms of subjective assessment of well-being, *in vivo* pharmacokinetic characteristics over 12 h, the adverse event (AE) profile and measurement of vital signs and conjunctival reddening over 12 h. Secondary objectives of GWPD9901 were to compare the effects of the four preparations in terms of cognitive assessment, subjective assessment of well-being *in vivo* pharmacokinetic characteristics over 12 h, the AE profile and measurement of vital signs and conjunctival reddening over 12 h.

INVESTIGATIONAL PLAN

Periods 1, 5 and 6 were open label, Periods 2 to 4 double blind. The study was a partially randomised crossover using single doses of THC and/or CBD or placebo. In Period 1, each subject received CBD:THC as a series of sublingual drops. In Periods 2 to 4, the High CBD, High THC and placebo were administered as a series of sublingual drops according to a pre-determined randomisation scheme. In Period 5, the aerosol test treatment was administered sublingually via a pressurised aerosol and in Period 6 the test treatment was administered as an inhaled dose via a nebuliser. There was a minimum washout period of six-days between each dose.

Blood samples were taken for plasma concentration analysis and blood pressure (BP) and pulse, cognitive testing (Periods 1 to 4 only) and PD effects were measured and recorded at pre-determined times during each study period.

Six healthy subjects (three male and three female) who complied with all the inclusion and exclusion criteria were required to complete the study in its entirety.

The CBD:THC sublingual drops were administered in Period 1 as the combination of CBD and THC was thought to be safest and allow assessment of the tolerability of the other test treatments. High CBD, High THC and placebo sublingual drops were then fully randomised to prevent period effect and this part of the study was also double blind to ensure no bias was introduced when recording AEs and other parameters.

The pressurised aerosol and inhaled nebuliser routes of administration were chosen to assess different methods of dose administration. These doses were not blinded or randomised due to the contrasting method of administration.

Subjects were admitted to the clinical unit the evening before dosing (Day− 1) to allow dietary control and eligibility assessments to be made. Dose administration was in the morning of Day 1 of each period to allow for measurements/assessments to be carried out up to 12 h post-dose with minimal disruption to the subjects sleep. A crossover design was chosen to enable both inter- and intra-subject comparisons of the data collated. A six-day washout period was chosen as it was estimated that plasma concentrations of cannabinoids would be below the Lower Limit of Quantification (LLOQ) before administration of the next dose and to facilitate scheduling within the clinical unit.

This was a proof of concept study and therefore a small number of subjects (six) were required.

INCLUSION CRITERIA

For inclusion in the study, subjects were required to fulfil all of the following criteria:

1. Were aged 30-45 years.
2. Weighed between 50-90 kg inclusive and body mass index (BMI) no greater than 30 kg/m^2.
3. Were willing and able to undertake all study requirements including pre- and post-study medical screening.
4. Had given written informed consent.
5. Female: were surgically sterilised or were taking adequate contraceptive precautions.
6. Male: agreed to use barrier methods of contraception both during and for three months after completing the study.
7. Were cannabis experienced but had abstained for a minimum of 30 days prior to receiving the first dose.

EXCLUSION CRITERIA

Subjects were deemed not acceptable for participation in the study if any of the following criteria applied:

1. Had evidence of clinically significant cardiovascular, haematological, hepatic, gastro-intestinal, renal, pulmonary, neurological or psychiatric disease.
2. Had a history of schizophrenic-type illness.
3. Had a history of chronic alcohol or drug abuse or any history of social drug abuse other than experience with cannabis.

4. Had a resting systolic blood pressure (SBP) greater than 140 mmHg or diastolic blood pressure (DBP) greater than 90 mmHg.
5. Had a history of sensitivity to cannabis or multiple allergies or drug sensitivities.
6. Had a history of asthma.
7. Were currently taking any medication including self-medication.
8. Had taken a regular course of medication within the four weeks prior to first test treatment administration.
9. Had taken any medication within the fourteen days prior to first test treatment administration except for vitamins (which were required to be discontinued at screening), or the occasional use of paracetamol or, for females only, contraceptive preparations.
10. Had been hospitalised for any reason within the twelve weeks prior to first test treatment administration.
11. Had lost or donated greater than 400 ml of blood in the twelve weeks prior to first test treatment administration.
12. Had participated in a clinical trial in the 12 weeks prior to first test treatment administration.
13. Smoked more than five cigarettes a day.
14. Consumed more than 21 units of alcohol per week (male) or 14 units (female).
15. Had positive results for Hepatitis B or C, or Human Immunodeficiency Virus (HIV) 1 or 2 screening.
16. Had clinically significant biochemistry, haematology or urinalysis results at screening.
17. Were pregnant or lactating (females).
18. Refused to use the designated contraceptive precautions (male or female).
19. Failed to pass the Hospital Depression and Anxiety Scale (HADS) (reference to the Cognitive Assessment tests).
20. Were found to be colour blind (Ishihara colour blind screening).

STUDY RESTRICTIONS

Subjects were required to abstain from the following for the duration of the study:

i. All foods and beverages containing caffeine and alcohol for 48h pre-each dose until the end of each confinement period;

    ii. Drinking more than 3 units (male) or 2 units (female) of alcohol per day during non-restricted days.

    iii. Taking any drugs, including drugs of abuse, prescribed and/or over-the-counter medications for four weeks prior to first dose and for the duration of the study.

REMOVAL OF SUBJECTS FROM THERAPY OR ASSESSMENT

The subjects were free to withdraw from the study without explanation at any time and without prejudice to future medical care. Subjects may have been withdrawn from the study at any time if it was considered to be in the best interest of the subject's safety.

TEST TREATMENTS ADMINISTERED

All subjects received a single dose of the allocated test treatment on Day 1 of each of the six periods. All subjects received five single doses (maximum of 20 mg CBD and/or THC per dose) of CBD and/or THC and one placebo dose. Preparations were as follows (Table 1).

Test treatments consisted of CBD:THC sublingual drops (GW-1011-01): 5 mg Δ^9-tetrahydrocannabinol (THC) + 5 mg cannabidiol (CBD) per ml of glycerol:ethanol (Eth):propylene glycol (PG) (4:4:2), with peppermint flavouring, High CBD sublingual drops (GW-3009-01): 5 mg CBD per ml of glycerol:Eth:PG (4:4:2), with peppermint flavouring, High THC sublingual drops (GW-2009-01): 5 mg THC per ml of glyc-

TABLE 1. Product Codes, Batch Numbers and Expiry Dates for Each Test Treatment

Treatment	Batch No.	Product Code	Expiry Date	Total Dose	No. of Drops/Sprays/ Inhalations
CBD:THC SL Drops	90903	GW-1011-01	Oct 31, 1999	20 mg CBD + 20 mg THC	8 (10 mins apart)
CBD SL Drops	90902	GW-3009-01	Oct 31, 1999	20 mg CBD	8 (10 mins apart)
THC SL Drops	90901	GW-2009-01	Oct 31, 1999	20 mg THC	8 (10 mins apart)
Aerosol	91001	GW-1009-01	Nov 21, 1999	20 mg CBD + 20 mg THC	8 (10 mins apart)
Nebuliser	91002	GW-1012-01	Oct 27, 1999	20 mg CBD + 20 mg THC	series of 50 breaths (5 mins apart)

SL = sublingual

erol:Eth:PG (4:4:2), with peppermint flavouring, placebo sublingual drops (GW-4003-01): glycerol:Eth:PG (4:4:2), with peppermint flavouring, aerosol (GW-1009-01): 5 mg CBD + 5 mg THC per ml formulated in propellant:Eth (80:20), and nebuliser (GW-1012-01): 10 mg CBD + 10 mg THC per ml of cremophor (Crem) (0.4):PG (1.5):macrogol (1):dodecanol (0.8):H_2O (7.4), and placebo nebuliser (administered to subjects 005 and 006 instead of the active nebuliser test treatment): Crem (0.4):PG (1.5):macrogol (1):dodecanol (0.8):H_2O (7.4).

Each test treatment container was identified with no less than study number, subject number, period number, unit number and expiry date. All subjects received CBD:THC sublingual drops in Period 1, the aerosol test treatment in Period 5 and the inhaled nebuliser test treatment in Period 6. High CBD, High THC and placebo sublingual drops were randomised in Periods 2 to 4 according to the randomisation scheme. The doses were chosen as they were considered to be the average dose of cannabinoids received by smoking a cannabis cigarette. Subjects were allowed to stop dosing at any time if effects were too unpleasant. The Principal Investigator was also permitted to stop dosing before the maximum of 20 mg CBD and/or 20 mg THC was achieved if it was considered that the PD effects were too great.

Subjects 005 and 006 received placebo via the nebuliser to determine if the adverse effects that subjects 001 to 004 had experienced were due to the method of administration or the active ingredient.

The test treatments were administered in the morning of each dosing day according to the randomisation scheme. Subjects were dosed in the morning to allow for measurements to be taken and procedures to be carried out to prevent the subjects being confined to the clinical unit overnight after dosing. A minimum of six-days washout between each dose was specified as it was considered that by that time, plasma cannabinoid concentrations would be below the LLOQ.

BLINDING

Periods 1, 5 and 6 were open label. Periods 2 to 4 were double blind. Unblinding envelopes were retained at the study centre and a duplicate set was retained at GW. All subjects completed the study without any serious adverse events (SAEs), therefore unblinding of any subject test treatment was not required. Upon completion of the in-life phase of the study, all unblinding envelopes were returned to GW intact.

Subjects were required to abstain from taking any medication in the 14 days, and/or taking a course of medication in the four weeks prior to the study commencing. Any medications taken by subjects during the study (screening to post-study examination) were recorded in the Case Report Form (CRF) and Investigator judgement as to the subjects' continued eligibility was made.

TREATMENT COMPLIANCE

Subjects were dosed by the Principal Investigator or suitably trained designee. For the sublingual drops and pressurised aerosol test treatments, subjects were instructed to allow each drop/spray to absorb under their tongue and not to swallow for as long as possible. For the nebuliser test treatment, subjects were instructed to breathe normally whilst inhaling through the nebuliser. The nebuliser was breath activated and subjects were instructed to inhale for 50 breaths over approximately 5 min, stop and repeat after 10 min. This process was required to be repeated until the maximum dose was reached or dosing was stopped. The actual time of administration of each drop/spray was recorded in the CRF and the dosing procedure was witnessed by a dose verifier. Due to a problem with the nebuliser, which did not give the required dose over 50 breaths, subjects were permitted to take more than 50 breaths per series.

PRE-STUDY SCREENING

Subjects were required to undergo a pre-study screen no more than 14 days prior to first dose administration to determine their eligibility to take part in the study. Only those subjects who were healthy and complied with all the study requirements were deemed eligible for participation.

Demographic Data

The subjects' date of birth, age, sex, race, height, weight, body mass index (BMI), previous cannabis experience, tobacco and alcohol consumption were recorded.

Concomitant Medications and Medical History

Subjects were asked to provide details of any drugs, vitamins or medications they had taken in the previous four weeks or were currently taking. If taking vitamins or paracetamol at screening, subjects were required to stop taking them at screening to be eligible for the study. Previous medical history details were also recorded.

Physical Examination

Subjects underwent a physical examination to determine if there are any abnormalities in any body systems. Blood pressure (systolic/diastolic) and pulse were measured after the subject had been seated for no less than 5 min. A 12-lead electrocardiogram (ECG) was taken for each subject and assessed using the usual parameters.

Microscopy was required to be carried out on any abnormal urine samples. A pregnancy test was carried out on all urine samples from female subjects. The samples provided (male and female) were also used to screen for drugs of abuse. A blood sample was taken in an EDTA blood tube for haematology. A blood sample was taken in a gel blood tube for clinical chemistry. A blood sample was taken in a gel blood tube to screen for the presence of Hepatitis B and/or C and/or HIV.

Subjects were required to complete the HADS test and the Ishihara Colour Blindness test.

PRE-DOSE PROCEDURES

The day before dosing for Period 1, subjects were required to attend the clinical unit in the afternoon to complete a baseline well-being questionnaire and cognitive assessment. In all other periods, subjects were required to arrive at the clinic at approximately 11 h prior to dosing (i.e., the previous evening). A snack was provided at approximately 21:00 and thereafter subjects were required to fast until 4 h post-dose. Subjects were required to complete the Adult Reading Test. On the morning of each dosing day, each subject's health status was updated and pre-dose procedures (blood pressure and pulse, alcohol and drug of abuse screen and pregnancy test for female subjects) were carried out.

Within the 30 min before dosing started the following pre-dose procedures were carried out: cardiac monitoring was started, blood pressure and pulse recorded, conjunctival reddening assessed and a well-being questionnaire completed. The pre-dose blood sample was also taken.

Blood Sampling for Plasma Concentration Analysis

Blood samples (5 ml) were collected into 5 ml lithium heparin blood tubes via indwelling cannula or individual venipuncture. Samples were placed immediately into an ice bath until centrifuged (1000 G for 10 min at 4°C). The resultant plasma was decanted into two identical pre-labelled amber glass plasma tubes and placed in a freezer at −20°C.

Blood samples were collected pre-dose and at 5, 10, 15, 30 and 45 min and at 1, 2, 3, 4, 6, 8 and 12 h post-dose.

Plasma Concentration Analytical Procedures

Plasma concentrations of CBD, THC and 11-hydroxy-THC were measured in each plasma sample according the analytical protocol.

SAFETY ASSESSMENTS

Urine Drug Screen

Urine drug screens were required to be carried out at check-in for each study period. The drug screen was required to be negative for all drugs pre-dose Period 1. In subsequent periods, positive THC results may have occurred due to administration of test treatment in the previous period and therefore screening for THC was not carried out. The urine sample was required to be negative for all other drugs tested for the subject to be eligible to continue.

Blood Pressure and Pulse

Subjects' blood pressure and pulse were measured pre-dose then at 5, 10, 15, 30 and 45 min and at 1, 2, 3, 4, 6, 8 and 12 h post-dose.

Cardiac Monitoring

Cardiac monitoring was carried out continually from pre-dose to 4 h post-dose for each subject. A print out from the monitor was retained with the study centre study files.

Conjunctival Redness

Subjects were visually assessed for conjunctival reddening at the following times: pre-dose, 15, 30 and 45 min and at 1, 2, 4, 8 and 12 h post-dose. The extent of reddening was scored according to Table 2.

TABLE 2. Conjunctival Reddening Guidelines

Condition	Score
No reddening apparent	0
Slight reddening	1
Moderate reddening	2
Severe reddening	3

Adverse Events

Subjects' health was monitored continuously throughout the study. Subjects were also encouraged to inform the clinical staff of any changes in their health as soon as possible. All AEs were recorded in the CRF and followed to resolution or at the discretion of the Investigator.

Cognitive Assessments

Cognitive tests were carried out in Periods 1 to 4 only, using the Cambridge Neuropsychological Test Automated Battery (CANTAB), supplied by CeNeS, Histon, Cambridgeshire, UK. Subjects were asked to complete cognitive tests on the day before dose in Period 1 (baseline), and in each period at 10 min post last actuation then at 3 and 8 h post-dose.

Well-Being Questionnaire

Subjects were required to complete a series of visual analogue scales for alertness, well-being, mood, dryness of mouth, hunger level and any unpleasant effects. These were carried out on Day−1 and then in each period pre-dose, 10 min and 3, 8 and 12 h post-dose.

FOOD AND BEVERAGES

Dietary Restrictions

The subjects were instructed not to consume alcohol or caffeine-containing food or drink from 48 h before dosing in each period until after they were discharged from the clinical unit. During treatment free days subjects were required to limit their alcohol intake to no more than three units per day (males) or two units per day (females).

A snack was provided at 21:00 on the evening before dosing, thereafter subjects were required to fast until 4 h post-dose. After 4 h post-dose, decaffeinated drinks were provided *ad libitum*. Lunch and dinner were provided at 4 and 9 h post-dose, respectively. Subjects were provided with breakfast prior to discharge at 24 h post-dose. With the exception of the snack at 13 h post-dose and breakfast on Day 2, subjects were required to eat the entire meals provided. The details of diet are presented in Table 3.

Check-Out Procedures

Following breakfast on Day 2 (approximately 24 h post-dose) of each study period and if deemed well enough to leave by the Investigator, subjects were discharged from the clinical unit. Prior to discharge, any ongoing AEs were updated and follow-up arranged if required.

Post-Study Screening

Each subject was required to return to the clinical unit no more than ten days post last dose to undergo a post-study examination. This consisted of a physical examination, blood samples taken for haematology and clinical chemistry, urinalysis, a 12-lead ECG and vital signs recorded. Any ongoing AEs were updated and, if required, arrangements were made to follow up with the subjects.

TABLE 3. Suggested Menu

Meal	Time	Content
Evening Meal	Day −1, 21:00	Two filled rolls
		One light desert (e.g., yoghurt)
		One piece of fruit
		Decaffeinated drink
Lunch	Day 1, 4 h post-dose	Cooked meal (e.g., meat and two vegetables)
		Dessert
Evening Snack	Day 1, 13 h post-dose	Optional, no restrictions
Breakfast	Day 2	Optional, no restrictions

DATA QUALITY ASSURANCE

Study Monitoring

All details regarding the study were documented within individual CRFs provided by GW for each subject. All data recorded during the study were checked against source data and for compliance with GCP, internal Standard Operating Procedures (SOPs), working practices and protocol requirements. Monitoring of the study progress and conduct was carried out by the Clinical Department of GW according to GW SOPs and was ongoing throughout the study.

Standardisation of Laboratory Procedures

Analysis of safety bloods (haematology and clinical chemistry) was carried out by Unilabs UK (previously J S Pathology Ltd).

Investigator Responsibilities

The Investigator was responsible for monitoring the study conduct to ensure that the rights of the subject were protected, the reported study data were accurate, complete and verifiable and that the conduct of the study was in compliance with ICH GCP. At the end of the study, the Principal Investigator reviewed and signed each CRF declaring the data to be true and accurate. If corrections were made after review the Investigator acknowledged the changes by re-signing the CRF.

Clinical Data Management

Data were double entered into approved data tables using Microsoft Excel software. Manual checks for missing data and inconsistencies were carried out and queries were raised for any resulting issues.

Once the data were clean, i.e., no outstanding queries, then Quality Control (QC) checks of 100% of the data for a 10% sample of the patients were conducted to make a decision on the acceptability of the data. Any errors were resolved and any error trends across all subjects were also corrected. Upon completion of the QC step, the data sets were burnt onto a compact disc.

Quality Assurance Audits

Clinical Quality Audits were carried out.

Statistical and Analytical Plans

The statistical analysis was carried out as indicated in the protocol. All statistical analyses were performed using SAS® for Windows (v8) software.

Pharmacokinetic Analysis

All p-values quoted are two-sided. No blood samples were missed in the subjects who were dosed therefore all subjects were deemed evaluable for and were included in pharmacokinetic analyses. The pharmacokinetic parameters calculated were as noted in Table 4.

Summary statistics were calculated for each pharmacokinetic parameter and treatment (arithmetic mean, number (N), standard deviation (SD), coefficient of variance (CV%), minimum and maximum for all parameters and additionally the geometric mean for AUC_{0-t}, $AUC_{0-\infty}$ and C_{max}). AUC_{0-t}, $AUC_{0-\infty}$ and C_{max} were natural log transformed prior to analysis and T_{max} was analysed untransformed; $t_{1/2}$ and K_{el} were summarised only. Each parameter was analysed using analysis of variance (ANOVA) with subject and treatment as factors. Least square (LS) means were presented for each test treatment. Point estimates (differences between least square means) for the contrasts between each of High THC, aerosol and inhaler with CBD:THC were presented with the corresponding 95% confidence intervals (CI); for the log-transformed variables, the contrasts were first back transformed to provide ratios and corresponding 95% confidence intervals. The distribution of T_{max} was also summarised.

TABLE 4. Pharmacokinetic Definitions

T_{max}	Time to the maximum measured plasma concentration.
C_{max}	Maximum measured plasma concentration over the time span specified.
$t_{1/2}$	Putative effective elimination half life (the initial descending portion of each plasma concentration-time graph).
AUC_{0-t}	The area under the plasma concentration versus time curve, from time zero to 't' (where t = the final time of positive detection, t ≤ 12h) as calculated by the linear trapezoidal method.
$AUC_{0-\infty}$	The area under the plasma concentration versus time curve from zero to t calculated as AUC_{0-t} plus the extrapolated amount from time t to infinity.
K_{el}	The elimination rate constant.

Pharmacodynamic Analysis

All subjects who completed at least one study period were evaluable for pharmacodynamic analysis. All pharmacodynamic parameters were summarised by test treatment group and analyte. Data for conjunctival reddening and well-being questionnaire were summarised descriptively by time point and treatment (arithmetic means, N, SDs, medians, minima and maxima or counts and percentages, as appropriate). The changes from pre-dosing for the well-being questionnaire were summarised similarly. Analysis of the cognitive assessments was carried out by CeNeS Ltd.

SAFETY ANALYSIS

Adverse Events

All AEs were coded by Medical Dictionary of Regulatory Activities (MedDRA) and presented by system organ class (SOC) and preferred term (PT). Laboratory data collected pre and post-study were summarised descriptively (N, mean, SD, median, minimum and maximum) at each of the two time-points and also as the change from pre-study to post-study.

Blood Pressure and Pulse

For blood pressure and pulse descriptive statistics (N, mean, SD, median, minimum and maximum) were calculated and summarised at each time point by treatment group. In addition, the calculations were performed for the absolute change in means from pre-dose. Blood pressure and pulse data are listed for each subject at each time point.

12-Lead ECG

For each of the ECG parameters (heart rate (HR), PR interval, QT interval and QRS width), descriptive statistics (N, mean, SD, median, minimum and maximum) were calculated and summarised pre- and post-study.

Determination of Sample Size

No formal sample size calculation was carried out for this study, as it was a "First in man" safety and tolerability study.

Changes in the Conduct of the Study or Planned Analyses

The protocol stated that the pharmacokinetic parameters AUC_{0-t}, C_{max}, C_{res} and T_{max} would be evaluated. In accordance with standard practice, T_{max}, C_{max}, AUC_{0-t}, $AUC_{0-\infty}$ were evaluated and compared between treatments. In addition, $t_{1/2}$ and K_{el} were summarised only.

Study Subjects

Three healthy male and three healthy female subjects were required to complete the study in its entirety. Six male and six female subjects were randomised and all of those subjects completed the study. No subjects withdrew from the study and no replacements were required. Only one minor protocol deviation was reported throughout the study. One subject consumed caffeine in the 48 h prior to dosing for Period 4. This was not considered by the Investigator to affect the subject's eligibility and is not considered to affect the integrity of the study.

Plasma Concentration and Pharmacokinetic Evaluation

Six healthy subjects (three male and three female) were required to complete the study in its entirety. Six subjects (001 to 006) who were randomised in the study were included in the data analysis.

Demographic and Baseline Characteristics

All subjects included in the study complied with all demographic and baseline requirements.

Measurements of Compliance

Each test treatment was administered by suitably trained clinical staff. No deviations to the dosing regimen were noted for any subject throughout the study.

INDIVIDUAL PLASMA CONCENTRATION DATA AND PHARMACOKINETIC RESULTS

Analysis of Plasma Concentration Results

Plasma samples were analysed for CBD, THC and 11-hydroxy-THC according to the analytical protocol. Analytical results were produced in tabular form and concentration-time graphs were produced from these data. Mean plasma concentrations are summarised in Table 5.

The LLOQ was 1 ng/ml. Data below the LLOQ are presented as <1 and the actual value measured is presented in parenthesis. The actual values measured were used when creating graphs.

CBD:THC Sublingual Drops

Mean concentrations of CBD, THC and 11-hydroxy-THC were above the LLOQ by 45 min post-dose (Figure 1) (range of individual times: 45-180 min, CBD; 30-120 min THC and 11-hydroxy-THC). Mean concentrations of THC (Table 6) were at least double those of CBD throughout the sampling period and from 120 min to the end of sampling mean concentrations of 11-hydroxy-THC were approximately double those of THC (CBD 1.23 ng/ml, THC 3.13 ng/ml, 11-hydroxy-THC 6.68 ng/ml). By 360 and 480 min post-dose the mean level of CBD and THC, respectively and all individual levels were below the LLOQ.

High CBD Sublingual Drops

Mean concentrations of CBD were above the LLOQ by 30 min post-dose (range: 30-120 min), peaked at 120 min (1.49 ng/ml) and

TABLE 5. Mean Plasma Concentration Data

Time (min)	Mean Plasma Concentrations											
	CBD				THC				11-Hydroxy THC			
	CBD: THC SL Drops	High CBD SL Drops	Aerosol	Nebuliser	CBD: THC SL Drops	High THC SL Drops	Aerosol	Nebuliser	CBD: THC SL Drops	High THC SL Drops	Aerosol	Nebuliser
0	0.00	0.00	0.00	0.00	0.00	0.00	0.00	0.00	0.00	0.00	0.00	0.00
5	0.00	0.00	0.00	6.81	0.00	0.00	0.00	9.45	0.00	0.00	0.00	0.00
10	0.00	0.00	0.00	3.26	0.00	0.00	0.00	5.88	0.00	0.00	0.00	0.26
15	0.00	0.00	0.24	5.04	0.00	0.21	0.23	6.50	0.00	0.00	0.00	0.22
30	0.00	0.33	0.48	5.40	0.19	1.03	1.00	8.25	0.34	1.17	0.72	0.43
45	0.60	0.58	0.96	2.91	1.64	1.71	1.49	4.44	1.70	2.57	1.68	0.24
60	1.20	0.93	0.97	4.56	3.04	3.33	1.87	6.74	3.51	4.36	2.56	0.55
120	1.64	1.49	0.73	0.96	4.67	3.86	2.38	2.31	7.43	5.19	4.84	0.29
180	1.23	0.73	0.86	0.39	3.13	2.94	2.36	0.91	6.68	4.81	5.07	0.20
240	0.48	0.45	0.60	0.29	1.70	1.34	1.38	0.36	4.82	3.14	3.90	0.00
360	0.00	0.00	0.99	0.00	0.55	0.00	1.30	0.00	2.43	1.01	2.89	0.00
480	0.00	0.00	0.22	0.00	0.00	0.00	0.22	0.00	1.05	0.23	1.40	0.00
720	0.00	0.00	0.00	0.00	0.00	0.00	0.00	0.00	0.20	0.00	0.17	0.00

FIGURE 1. GWPD9901: Mean Plasma Cannabinoid Concentrations Following Adminstration of CBD:THC, 1:1 Sublingual Drops

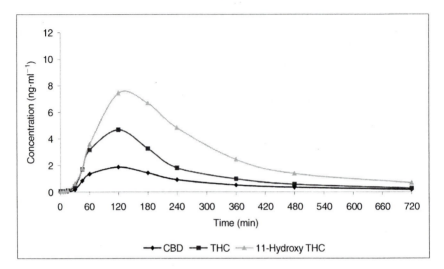

TABLE 6. Mean Pharmacokinetic Parameters

Time (min)	Mean Pharmacokinetic Parameters				
	T_{max} (min)	C_{max} (ng/ml)	AUC_{0-t} (ng/ml.min)	$t_{1/2}$ (min)	$AUC_{0-\infty}$ (ng/ml.min)
CBD					
CBD:THC SL Drops	100	2.58	209.30	118.33	578.89
High CBD SL Drops	130	2.05	156.13	NC	NC
Aerosol	141	2.60	325.93	143.77	811.75
Nebuliser	36	9.49	564.35	65.71	726.81
THC					
CBD:THC SL Drops	100	6.50	737.48	78.53	928.42
High THC SL Drops	110	5.77	628.80	65.53	818.10
Aerosol	130	3.69	636.11	83.00	776.09
Nebuliser	32	12.46	786.33	47.13	899.77
11-Hydroxy THC					
CBD:THC SL Drops	140	8.25	1842.75	117.68	2066.30
High THC SL Drops	110	7.29	1163.78	99.55	1373.19
Aerosol	160	6.23	1568.20	138.11	1838.04
Nebuliser	38	1.65	65.15	132.56	495.67

NC = Not acceptable

thereafter declined such that they were below the LLOQ by 360 min in all subjects (Figure 2). Mean plasma concentrations of CBD were generally similar to those seen for CBD:THC sublingual drops (Table 6). Neither THC nor 11-hydroxy-THC was detected in quantifiable amounts throughout the sampling period.

High THC Sublingual Drops

Mean concentrations of THC were above the LLOQ by 15 min post-dose (individual range: 15-60 min) (Figure 3), which was marginally earlier than for the CBD:THC sublingual drops (45 min post-dose). Mean concentration reached a peak around 120 min (3.86 ng/ml) (Table 6) and by 360 min had declined below the LLOQ. Mean concentrations of 11-hydroxy-THC were above the LLOQ by 30 min post-dose (individual range: 30-60 min) (Figure 3), which was also marginally earlier than for the CBD:THC sublingual drops (45 min post-dose). Mean concentration reached a peak around 120 min (5.19 ng/ml) and by 480 min had declined below the LLOQ. Concentrations of THC and 11-hydroxy-THC were generally similar to those seen after the CBD:THC sublingual drops.

FIGURE 2. GWPD9901: Mean Plasma Cannabinoid Concentrations Following Administration of High CBD Sublingual Drops

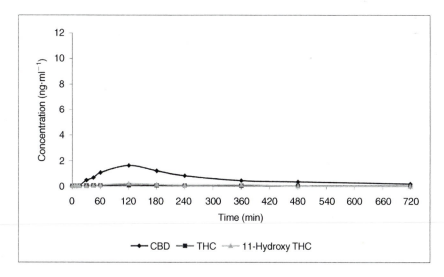

FIGURE 3. GWPD9901: Mean Plasma Cannabinoid Concentrations Following Administration of High THC Sublingual Drops

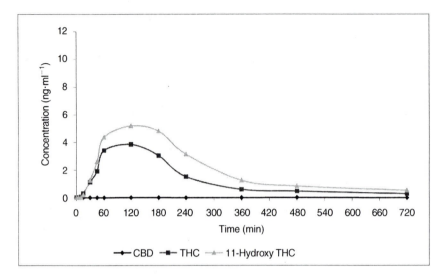

Placebo Sublingual Drops

Following placebo dosing no quantifiable amount of any cannabinoid was detected in any subject during the sampling period.

Pressurised Aerosol

Mean concentrations of CBD and THC above the LLOQ were detected in plasma by 15 min post-dose (range 10-180 min for CBD (excepting Subject 006 for whom concentrations remained below LLOQ); 15-180 min for THC) which was marginally earlier than for the CBD:THC sublingual drops (Figure 4).

Mean concentrations of CBD show two similar peak levels at 60 and 360 min (0.97 and 0.99 ng/ml, respectively) (Figure 4) reflecting the variability in the time of peak plasma concentration (range 45-360 min) between individuals. Mean concentrations of CBD had declined below LLOQ by 720 min.

Mean concentrations of THC peaked around 120-180 min (2.38 ng/ml, 2.36 ng/ml) and had declined below LLOQ by 720 min (Figure 4). Mean concentrations of 11-hydroxy-THC above the LLOQ were de-

FIGURE 4. GWPD9901 Extension: Mean Plasma Cannabinoid Concentrations
Following Administration of CBD:THC 1:1 Aerosol

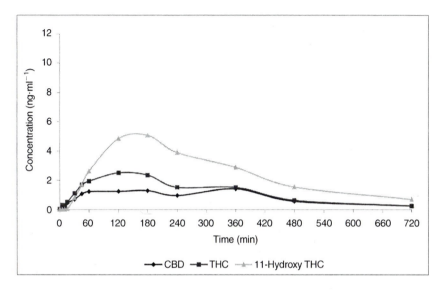

tected in plasma by 30 min post-dose (range: 30-120 min), peaked
around 180 min (5.07 ng/ml) and then declined more slowly than THC
or CBD and remained above the LLOQ at 720 min (Figure 4).

Mean concentrations of THC were generally greater than those for
CBD but less than mean concentrations of 11-hydroxy-THC (Table 6).
Mean concentrations of CBD, THC and 11-hydroxy-THC following
the pressurised aerosol were generally higher than for the CBD:THC
sublingual drops from 45-60 min to 240 min and were lower than for the
CBD:THC sublingual drops at almost all other time points. At 360 min
to 720 min post-dose mean concentrations of each cannabinoid were
marginally greater for the pressurised aerosol than for the CBD:THC
sublingual drops.

Following administration of the test treatment via the pressurised
aerosol, mean concentrations of each cannabinoid in plasma were above
the LLOQ for longer when compared to the CBD:THC sublingual drops.

Inhaled Nebuliser

The dose administered via the inhaled nebuliser was approximately
half that of the sublingual drops and aerosol. Mean concentrations of

CBD and THC were above the LLOQ by 5 min post-dose (range 5-30 min for both CBD and THC) and each cannabinoid was detected in plasma notably earlier than the CBD:THC sublingual drops (Figure 5).

Mean concentrations of CBD fluctuated considerably between 5 min and 60 min post-dose, reflecting the variability in levels and timing of peak concentrations in individuals, but were considerably higher than following the other treatments.

Mean concentrations of THC were higher than corresponding concentrations of CBD and also fluctuated considerably between 5 min and 60 min post-dose, reflecting the individual variability.

Mean concentrations of both CBD and THC declined rapidly from 60 min. CBD concentrations were below the LLOQ in all but one subject at 180 min and THC in all but one subject by 240 min.

Mean concentrations of 11-hydroxy-THC were much lower than corresponding concentrations of both CBD and THC and much less than following the other treatments (Figure 5). In three subjects, levels of 11-hydroxy-THC failed to rise above the LLOQ at all during the sampling period.

FIGURE 5. GWPD9901 Extension: Mean Plasma Cannabinoid Concentrations Following Administration of CBD:THC, 1:1 Nebuliser

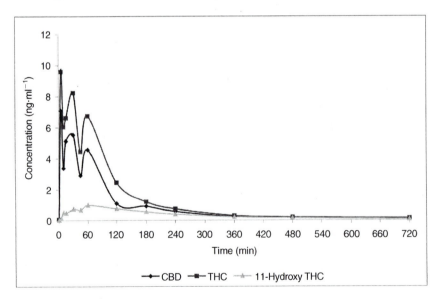

Analysis of Pharmacokinetic Parameters

PK parameters were calculated using WinNonlin® Professional 3.1. The model used was a non-compartmental, linear trapezoidal analysis. Values below the LLOQ were not used when calculating PK parameters. Mean values are presented in (Table 6).

The PK parameters for each test treatment (with the exception of placebo) were statistically compared to the PK parameters for the CBD:THC sublingual drops. Due to the low concentrations of cannabinoids in plasma some individual PK parameters were not calculable and therefore some of the mean PK parameters are not based on all six subjects.

CBD:THC Sublingual Drops

Following the CBD:THC sublingual drops arithmetic mean T_{max} of CBD (Table 7) and THC (Table 8) was 100 and 100 min, respectively. Arithmetic mean C_{max} of CBD was 2.58 ng/ml, arithmetic mean AUC_{0-t} 209.3 ng/ml.min and $AUC_{0-\infty}$ 578.89 ng/ml.min. The corresponding values for THC were greater as C_{max} was 6.50 ng/ml, AUC_{0-t} 737.48 ng/ml.min and $AUC_{0-\infty}$ 928.42 ng/ml.min. The arithmetic mean T_{max} of 11-hydroxy-THC was 140 min (Table 9). Arithmetic mean C_{max} was 8.25 ng/ml, arithmetic mean AUC_{0-t} 1842.75 ng/ml.min and $AUC_{0-\infty}$ 2066.30 ng/ml.min.

High CBD Sublingual Drops

The mean PK parameters for CBD following administration of High CBD sublingual drops were not statistically significantly different from

TABLE 7. CBD Pharmacokinetic Parameters: CBD:THC Sublingual Drops

Subject	T_{max} (min)	C_{max} (ng/ml)	AUC_{0-t} (min*ng/ml)	K_{el} (1/min)	$t_{1/2}$ (min)	$AUC_{0-\infty}$ (min*ng/ml)
1	120	3.70	264.00	NC	NC	NC
2	60	2.63	449.18	0.0048	144.57	749.51
3	60	1.95	14.63	NC	NC	NC
4	180	2.75	208.50	NC	NC	NC
5	60	2.64	266.10	0.0075	92.10	408.27
6	120	1.78	53.40	NC	NC	NC
Mean	100	2.58	209.30	0.0062	118.33	578.89
SD**	60-180	0.68	158.72	0.0019	37.10	241.29

** T_{max} presented as minimum-maximum
NC = Not calculable

TABLE 8. THC Pharmacokinetic Parameters: CBD:THC Sublingual Drops

Subject	T_{max} (min)	C_{max} (ng/ml)	AUC_{0-t} (min*ng/ml)	K_{el} (1/min)	$t_{1/2}$ (min)	$AUC_{0-\infty}$ (min*ng/ml)
1	120	9.29	880.28	0.0157	44.28	971.00
2	60	6.44	1005.75	0.0076	90.62	1327.35
3	60	5.62	287.85	0.0218	31.80	357.60
4	180	6.31	916.20	0.0077	89.71	1118.09
5	60	4.93	549.98	0.0083	83.77	686.54
6	120	6.38	784.80	0.0053	131.02	1109.93
Mean	100	6.50	737.48	0.0111	78.53	928.42
SD**	60-180	1.49	269.76	0.0063	35.81	350.49

** T_{max} presented as minimum-maximum
NC = Not calculable

TABLE 9. 11-Hydroxy THC Pharmacokinetic Parameters: CBD:THC Sublingual Drops

Subject	T_{max} (min)	C_{max} (ng/ml)	AUC_{0-t} (min*ng/ml)	K_{el} (1/min)	$t_{1/2}$ (min)	$AUC_{0-\infty}$ (min*ng/ml)
1	120	12.14	2459.40	0.0089	78.21	2586.89
2	180	7.87	1996.05	0.0060	115.20	2205.45
3	60	6.52	1208.85	0.0075	92.51	1343.65
4	240	7.22	2371.80	0.0045	153.34	2635.06
5	120	6.95	1459.20	0.0057	122.16	1758.81
6	120	8.82	1561.20	0.0048	144.64	1867.94
Mean	140	8.25	1842.75	0.0062	117.68	2066.30
SD**	60-240	2.07	512.22	0.0017	29.04	503.98

** T_{max} presented as minimum-maximum
NC = Not calculable

the CBD:THC sublingual drops (Table 10). The arithmetic mean T_{max} was 130 min and arithmetic mean C_{max} 2.05 ng/ml, arithmetic mean AUC_{0-t} was numerically lower than following the CBD:THC sublingual drops at 156.13 ng/ml.min. $AUC_{0-\infty}$ was not calculable as there were generally few time points in any subjects when plasma levels of CBD exceeded the LLOQ (at a single sampling time in three subjects).

High THC Sublingual Drops

Only mean $AUC_{0-\infty}$ for 11-hydroxy-THC following administration of the High THC sublingual drops (Table 11) was statistically signifi-

TABLE 10. CBD Pharmacokinetic Parameters: High CBD Sublingual Drops

Subject	T_{max} (min)	C_{max} (ng/ml)	AUC_{0-t} (min*ng/ml)	K_{el} (1/min)	$t_{1/2}$ (min)	$AUC_{0-\infty}$ (min*ng/ml)
1	240	1.32	185.40	NC	NC	NC
2	120	3.17	563.03	0.0135	51.18	664.93
3	120	1.90	57.00	NC	NC	NC
4	60	3.21	49.73	NC	NC	NC
5	120	1.14	34.20	NC	NC	NC
6	120	1.58	47.40	NC	NC	NC
Mean	130	2.05	156.13	NC	NC	NC
SD**	60-240	0.92	207.01	NC	NC	NC

** T_{max} presented as minimum-maximum
NC = Not calculable

TABLE 11. 11-Hydroxy THC Pharmacokinetic Parameters: High THC Sublingual Drops

Subject	T_{max} (min)	C_{max} (ng/ml)	AUC_{0-t} (min*ng/ml)	K_{el} (1/min)	$t_{1/2}$ (min)	$AUC_{0-\infty}$ (min*ng/ml)
1	120	6.74	1105.05	0.0070	98.38	1261.17
2	120	7.97	1793.18	0.0072	96.32	1983.56
3	60	6.21	669.83	0.0055	126.84	944.30
4	180	7.83	1292.70	0.0091	76.32	1456.75
5	120	6.67	1205.63	0.0065	105.95	1441.02
6	60	8.31	916.28	0.0074	93.49	1152.32
Mean	110	7.29	1163.78	0.0071	99.55	1373.19
SD**	60-180	0.85	380.32	0.0012	16.58	354.80

** T_{max} presented as minimum-maximum
NC = Not calculable

cantly lower when compared to the CBD:THC sublingual drops (1373.19 ng/ml.min vs. 2066.30 ng/ml.min, p = 0.0358). For THC (Table 12), T_{max} was 110 min, C_{max} 5.77 ng/ml, AUC_{0-t} 628.80 ng/ml and $AUC_{0-\infty}$ 818.10 ng/ml. C_{max}, AUC_{0-t} and $AUC_{0-\infty}$ were slightly lower than following the CBD:THC sublingual drops and T_{max} was slightly later but these differences were not statistically significant.

Pressurised Aerosol

There were no statistically significant differences in PK parameters for CBD (Table 13), THC (Table 14) or 11-hydroxy-THC (Table 15)

TABLE 12. THC Pharmacokinetic Parameters: High THC Sublingual Drops

Subject	T_{max} (min)	C_{max} (ng/ml)	AUC_{0-t} (min*ng/ml)	K_{el} (1/min)	$t_{1/2}$ (min)	$AUC_{0-\infty}$ (min*ng/ml)
1	120	5.46	708.45	0.0083	83.65	952.23
2	120	4.84	747.68	0.0047	146.19	1325.57
3	60	4.25	262.68	0.0238	29.14	305.56
4	180	7.75	937.20	0.0226	30.70	1025.79
5	120	6.76	727.95	0.0139	49.75	819.10
6	60	5.55	388.88	0.0129	53.72	480.33
Mean	110	5.77	628.80	0.0144	65.53	818.10
SD**	60-120	1.28	251.80	0.0076	44.18	372.95

** T_{max} presented as minimum-maximum
NC = Not calculable

TABLE 13. CBD Pharmacokinetic Parameters: Pressurised Aerosol

Subject	T_{max} (min)	C_{max} (ng/ml)	AUC_{0-t} (min*ng/ml)	K_{el} (1/min)	$t_{1/2}$ (min)	$AUC_{0-\infty}$ (min*ng/ml)
1	45	1.38	179.85	NC	NC	NC
2	120	2.76	536.85	0.0082	84.39	665.90
3	60	2.39	81.18	NC	NC	NC
4	360	4.85	940.80	0.0108	63.92	1062.52
5	120	1.61	216.90	0.0024	283.00	706.84
6	NC	NC	0.00	NC	NC	NC
Mean	141	2.60	325.93	0.0071	143.77	811.75
SD**	45-360	1.38	352.68	0.0043	121.01	218.13

** T_{max} presented as minimum-maximum
NC = Not calculable

TABLE 14. THC Pharmacokinetic Parameters: Pressurised Aerosol

Subject	T_{max} (min)	C_{max} (ng/ml)	AUC_{0-t} (min*ng/ml)	K_{el} (1/min)	$t_{1/2}$ (min)	$AUC_{0-\infty}$ (min*ng/ml)
1	180	2.43	500.10	0.0143	48.45	570.00
2	120	4.66	814.95	0.0152	45.55	888.55
3	60	3.72	356.60	0.0101	68.78	466.74
4	180	4.64	1139.40	0.0105	66.28	1263.71
5	120	3.04	377.03	0.0068	101.54	573.32
6	120	3.67	628.58	0.0041	167.41	894.25
Mean	130	3.69	636.11	0.0102	83.00	776.09
SD**	60-180	0.88	299.70	0.0043	45.93	297.88

** T_{max} presented as minimum-maximum
NC = Not calculable

TABLE 15. 11-Hydroxy THC Pharmacokinetic Parameters: Pressurised Aerosol

Subject	T_{max} (min)	C_{max} (ng/ml)	AUC_{0-t} (min*ng/ml)	K_{el} (1/min)	$t_{1/2}$ (min)	$AUC_{0-\infty}$ (min*ng/ml)
1	180	4.32	820.43	0.0107	65.08	958.44
2	120	6.08	1380.68	0.0046	149.66	1639.78
3	60	6.03	1726.43	0.0046	149.78	2082.97
4	360	6.54	2290.50	0.0050	137.48	2490.83
5	120	7.10	1316.03	0.0043	160.82	1550.36
6	120	7.30	1875.15	0.0042	165.86	2305.87
Mean	160	6.23	1568.20	0.0056	138.11	1838.04
SD**	60-360	1.07	509.69	0.0025	37.12	565.82

** T_{max} presented as minimum-maximum
NC = Not calculable

between the pressurised aerosol and CBD:THC sublingual drops. Following dosing with the pressurised aerosol T_{max} of both CBD and THC were a little later than after dosing with the CBD:THC sublingual drops (CBD 141 vs. 100 min and THC 130 vs. 100 min). CBD C_{max}, was very similar but AUC_{0-t} and $AUC_{0-\infty}$ were greater than following CBD:THC sublingual drops whereas THC $C_{max,}$ AUC_{0-t} and $AUC_{0-\infty}$ were all less than the sublingual drops. None of these differences was statistically significant.

Inhaled Nebuliser

The dose administered via the inhaled nebuliser was approximately half that of the sublingual drops and aerosol. T_{max}, of both CBD (36 min) (Table 16) and THC (32 min) (Table 17) were much earlier than the corresponding values after the sublingual drops (100 min and 100 min, respectively) or aerosol (T_{max} THC = 130 min, CBD = 141 min), though only the difference in THC T_{max} was significant for sublingual drops (p = 0.0046). Mean C_{max} of CBD (9.49 ng/ml) was statistically significantly greater than for the CBD:THC sublingual drops (2.58 ng/ml) (p = 0.0104). Mean C_{max} of THC was similarly greater (12.46 vs. 6.5 ng/ml) though the difference was not statistically significant. Mean AUC_{0-t} of CBD (564.35 ng/ml.min) was greater than for the CBD:THC sublingual drops (209.30 ng/ml.min); however, with a p-value of 0.0529, this was not a statistically significant difference. The AUC_{0-t} and $AUC_{0-\infty}$ values for CBD following dosing with the inhaled nebuliser were greater, though not statistically significantly, than the correspond-

TABLE 16. CBD Pharmacokinetic Parameters: Nebuliser

Subject	T_{max} (min)	C_{max} (ng/ml)	AUC_{0-t} (min*ng/ml)	K_{el} (1/min)	$t_{1/2}$ (min)	$AUC_{0-\infty}$ (min*ng/ml)
1	5	22.62	442.55	0.0275	25.23	479.68
2	30	9.87	499.53	0.0166	41.70	587.36
3	45	2.74	219.60	0.0039	179.66	750.96
4	15	4.70	171.20	0.0278	24.92	235.20
5	60	14.75	1859.90	0.0124	55.69	1997.28
6	60	2.25	193.33	0.0103	67.05	310.36
Mean	36	9.49	564.35	0.0164	65.71	726.81
SD**	5-60	8.01	649.39	0.0096	58.25	649.66

** T_{max} presented as minimum-maximum
NC = Not calculable

TABLE 17. THC Pharmacokinetic Parameters: Nebuliser

Subject	T_{max} (min)	C_{max} (ng/ml)	AUC_{0-t} (min*ng/ml)	K_{el} (1/min)	$t_{1/2}$ (min)	$AUC_{0-\infty}$ (min*ng/ml)
1	5	25.85	633.30	0.0334	20.74	675.50
2	30	14.74	884.58	0.0144	48.18	963.81
3	60	3.73	449.63	0.0089	77.77	593.24
4	30	5.15	217.35	0.0191	36.21	368.84
5	5	21.97	2256.68	0.0164	42.21	2387.60
6	60	3.29	276.48	0.0120	57.69	409.64
Mean	32	12.46	786.33	0.0174	47.13	899.77
SD**	5-60	9.89	760.53	0.0086	19.45	759.49

** T_{max} presented as minimum-maximum
NC = Not calculable

ing values for CBD:THC sublingual drops. $AUC_{0-\infty}$, AUC_{0-t}, C_{max} and T_{max} for 11-hydroxy-THC (495.67 ng/ml.min, 65.15 ng/ml.min, 1.65 ng/ml and 38 min, respectively) (Table 18) were statistically significantly lower when compared to the CBD:THC sublingual drops (2066.30 ng/ml.min, 1842.75 ng/ml.min, 8.25 ng/ml and 140 min, respectively). The p-values were 0.0034, < 0.0001, < 0.0001 and 0.0054, respectively.

Analysis of Cognitive Assessments and Well-Being

For each test treatment period, subjects were required to undertake a battery of cognitive assessments (Periods 1 to 4 only) and complete a well-being questionnaire. Subjects were also required to report a series

TABLE 18. 11-Hydroxy THC Pharmacokinetic Parameters: Nebuliser

Subject	T_{max} (min)	C_{max} (ng/ml)	AUC_{0-t} (min*ng/ml)	K_{el} (1/min)	$t_{1/2}$ (min)	$AUC_{0-\infty}$ (min*ng/ml)
1	10	1.56	30.10	NC	NC	NC
2	45	1.18	25.65	NC	NC	NC
3	NC	NC	0.00	NC	NC	NC
4	NC	NC	0.00	NC	NC	NC
5	60	2.21	270.00	0.0052	132.56	495.67
6	NC	NC	NC	NC	NC	NC
Mean	38	1.65	65.15	0.0052	132.56	495.67
SD**	10-60	0.52	115.37	NC	NC	NC

** T_{max} presented as minimum-maximum
NC = Not calculable

of well-being parameters using visual analogue scales (VAS). Each pre-dose assessment was taken to be the baseline measurement for each well-being parameter for each period.

Wakefulness was rated (0 = very drowsy and 100 = fully alert). CBD:THC sublingual drops resulted in the greatest drop in feeling of wakefulness with a decrease in wakefulness of −32.5 from baseline (88.5) at 3 h post-dose. All other test treatments, with the exception of placebo, which showed increased wakefulness throughout, also showed the greatest effect on wakefulness at 3 h post-dose with a range of decreases of −11.5 (High CBD) to −20.2 (aerosol).

Well-being was rated (0 = feel terrible and 100 = feel wonderful). Each test treatment resulted in a reduction similar to that for placebo in feeling of well-being. The greatest reduction (14.2 relative to baseline (94.0) at 3 h post-dose) was as a result of the CBD:THC sublingual drops. High CBD resulted in a later (−4.0 relative to baseline at 8 h post-dose) maximum mean decrease and the aerosol test treatment resulted in an earlier (−2.3 relative to baseline at 10 min post-dose) maximum mean decrease.

Mood was rated (0 = feel terrible and 100 = feel wonderful). All test treatments resulted in a maximum mean decrease (−2.2 to −11.3 relative to baseline) in mood at 3 h post-dose with the exception of the aerosol test treatment which showed a maximum mean decrease (2.7 relative to baseline) at 8 h post-dose.

Dry mouth was rated (0 = very dry and 100 = normal moisture). All test treatments, with the exception of the inhaler, resulted in a maximum mean increase in reporting of dry mouth at 3 h post-dose. CBD:THC sublingual drops resulted in the greatest increase in dryness of mouth with a maximum mean of −48.7 relative to baseline. The nebuliser test treatment resulted in an earlier maximum mean decrease with a change of −10.8 (relative to baseline) at 10 min post-dose.

Hunger was rated (0 = very hungry and 100 = not hungry). All test treatments and placebo resulted in a maximum mean increase in reported feeling of hunger at 3 h post-dose. The range of change from baseline was −9.8 (CBD:THC sublingual drops) to −27.5 (aerosol). The maximum mean increase in hunger following the placebo dose was −16.0.

Unpleasant effects were rated (0 = very unpleasant effects and 100 = no unpleasant effects). The maximum mean (relative to baseline) reporting of unpleasant effects was varied for each test treatment. CBD:THC sublingual drops was −11.5 at 3 h post-dose, High CBD was −9.3 at 12 h post-dose, High THC was −4.5 at 10 min post-dose, the aerosol was −5.7 at 8 h post-dose and the nebuliser was −14.0 at 10 min post-dose. The placebo treatment also resulted in reporting of unpleasant effects with a maximum mean increase of −6.5 at 10 min post-dose.

Post-Study Questionnaire

The results of the post-study questionnaire were assessed descriptively using frequency tables for each treatment. All of the subjects reported that the test treatment liked best was the 'liquid under the tongue' and the least liked test treatment was the nebuliser. Half of the subjects reported that the CBD:THC sublingual drops had the most pleasant effects and 67% (4) of the subjects reported that the nebuliser had the least pleasant effects. All of the subjects reported coughing and half reported a sore throat after administration of the test treatment via the nebuliser.

Analysis of Safety Parameters

The output from the cardiac monitors was intended for use at the clinical unit as an ongoing assessment for each subject. No concerns were raised as a result of the cardiac monitoring.

Pre-dose, all test treatments, including placebo but with the exception of High CBD, had between one and three subjects (17-50%) reported as having slight conjunctival reddening. Pre-dose High CBD, no conjunctival reddening was reported. Post-dose for all test treatments including placebo, the majority of subjects (67-100%) were reported as having 'slight' or 'no' conjunctival reddening. With the exception of

High CBD, moderate conjunctival reddening was reported in a maximum of two subjects (33% between 31 min and 4 h 01 min) for all test treatments including placebo. Only one subject had severe conjunctival reddening (on CBD:THC sublingual drops at 4 h 01 min).

Blood Pressure and Pulse During Treatment Periods

For each of the BP and pulse, parameters descriptive statistics (N, mean, SD, median, minimum and maximum) and the changes from pre-dose baseline were presented at each time point by test treatment group. In addition, the summaries were assessed for the absolute change from pre-dose.

12-Lead ECG

The ECG assessments (normal/abnormal) were assessed pre- and post-study.

Drug Dose, Drug Concentration and Relationships to Response

Each subject received three single doses of CBD:THC (20 mg CBD + 20 mg THC), one single dose of High THC (20 mg THC) and one single dose of High CBD (20 mg High CBD) (Table 19). The maximum total dose that was planned to be administered in the study was 80 mg CBD and 80 mg THC.

Drug-Drug and Drug-Disease Interactions

This study was carried out in healthy subjects who were not taking any medication.

Plasma Concentration Conclusions

Sublingual Drops

Following co-administration of CBD and THC as sublingual drops, mean concentrations of CBD, THC and 11-hydroxy-THC were above the LLOQ by 45 min post-dose. Plasma concentrations of THC were at least double those of CBD before both decreased below the LLOQ by 360 min and 480 min post-dose, respectively.

When High CBD sublingual drops were administered, plasma levels of CBD were generally similar to those measured after CBD:THC sublingual drops.

TABLE 19. Total Dose of Test Treatment Administered to Each Subject

Subject	Test Treatment					
	CBD:THC SL Drops	Placebo	High CBD SL Drops	High THC SL Drops	Aerosol	Nebuliser
1	20 mg CBD + 20 mg THC	0 mg	20 mg CBD	20 mg THC	20 mg CBD + 20 mg THC	2.5 mg CBD + 2.5 mg THC
2	20 mg CBD + 20 mg THC	0 mg	20 mg CBD	20 mg THC	20 mg CBD + 20 mg THC	2.5 mg CBD + 2.5 mg THC
3	12.5 mg CBD + 12.5 mg THC	0 mg	20 mg CBD	15 mg THC	20 mg CBD + 20 mg THC	2.5 mg CBD + 2.5 mg THC
4	20 mg CBD + 20 mg THC	0 mg	20 mg CBD	20 mg THC	20 mg CBD + 20 mg THC	2.5 mg CBD + 2.5 mg THC
5	20 mg CBD + 20 mg THC	0 mg	20 mg CBD	20 mg THC	20 mg CBD + 20 mg THC	0 mg**
6	20 mg CBD + 20 mg THC	0 mg	20 mg CBD	20 mg THC	20 mg CBD + 20 mg THC	0 mg**

** Subjects 005 and 006 received a placebo dose via the nebuliser

High THC resulted in mean levels of both THC and 11-hydroxy-THC being above the LLOQ earlier and also resulted in a slightly earlier decline than for CBD:THC. However, the concentrations of THC and 11-hydroxy-THC in plasma were similar or a little lower.

Pressurised Aerosol

Following administration of CBD:THC via the pressurised aerosol, mean levels of CBD and THC above the LLOQ were detected a little earlier than for the CBD:THC sublingual drops and declined below the LLOQ a little later. Plasma concentrations of THC, 11-hydroxy-THC and CBD were lower than following the sublingual drops.

Nebuliser

Following a dose administration of CBME via the nebuliser of approximately half that of the sublingual drops, mean plasma levels of both CBD and THC rose rapidly (within 5 min) to levels much higher than measured following sublingual drops and were maintained until around 120 min post-dose before declining rapidly. Levels of 11-hydroxy-THC were very low compared with those after sublingual dosing.

In conclusion, following sublingual administrations of CBD alone, THC alone or CBD:THC combined there was little difference in the plasma concentrations of THC or CBD. However, plasma levels of

CBD are less than corresponding levels of THC suggesting lower bioavailability. Following administration of CBD and THC by pressurised aerosol blood levels of both THC and CBD were lower compared with the sublingual drops. Following administration of CBD and THC via the nebuliser, there was rapid absorption and much greater plasma levels of both CBD and THC compared with sublingual dosing and the low levels of 11-hydroxy-THC suggests that metabolism of THC was significantly reduced.

Pharmacokinetic Conclusions

There were no statistically significant differences in the PK of THC or CBD between CBD:THC sublingual drops and High THC, High CBD or pressurised aerosol. With the exception of a single statistically significant difference in $AUC_{0-\infty}$ for 11-hydroxy-THC following administration of the High THC compared with CBD:THC sublingual drops there were no significant differences in the PK of 11-hydroxy-THC either. The differences in plasma concentrations and mean PK parameters observed between some of these treatments in the study were small relative to the individual variability.

Dosing with the inhaled nebuliser produced marked differences in the PK of CBD and THC compared with CBD:THC sublingual dosing. Peak concentration was greater and much earlier although only C_{max} of CBD and T_{max} of THC were statistically significantly different. Peak concentration and AUCs of 11-hydroxy-THC were statistically significantly less, reflecting reduced early metabolism of THC by this route.

In conclusion, no consistent statistically significant differences were noted between the PK parameters of High CBD, High THC and the aerosol when compared to the CBD:THC sublingual drops. However, the nebuliser resulted in a rapid absorption of CBD and THC and higher peak plasma levels but a reduction in the metabolism of THC to 11-hydroxy-THC.

Well-Being Conclusions

Results indicate that subjects experienced changes in wakefulness, feeling of well-being, mood, production of saliva and increased hunger and unpleasant effect. These were not clinically different following administration of each test treatment or placebo. The maximum mean reduction in wakefulness, feeling of well-being, mood and production of saliva were reported at 3 h post-dose and were as a result of CBD:THC sublingual drops.

Only small insignificant changes in wakefulness, feeling of well-being and mood were reported following administration of the placebo test treatment. However, a similar decrease in production of saliva, increase in hunger and marginally smaller incidence of unpleasant effects were seen with CBD:THC sublingual drops.

The greatest mean increase in hunger was reported following administration of the aerosol test treatment at 3 h post-dose. However, a similar effect was also observed at 3 h post-dose following administration of the placebo test treatment.

The greatest mean incidence of unpleasant effects was reported earlier than for any other effect and following administration of the nebuliser test treatment.

In conclusion, the decrease in general feeling of well-being were greatest following administration of CBD:THC sublingual drops.

Post-Study Questionnaire Conclusions

The sublingual test treatments were best liked and the nebuliser test treatment was least liked. The effects experienced following test treatment administration via the nebuliser were least liked. All of the subjects reported coughing and three subjects reported a sore throat following dosing during dosing with the nebuliser.

SAFETY EVALUATION

All six subjects completed all six periods of study treatment. The actual doses administered are presented in Table 19.

ADVERSE EVENTS

Brief Summary of Adverse Events

All six subjects experienced at least two AEs each during the study (Table 20). All the AEs were non-serious and most (32 events) were related to the study treatment. The majority of AEs were mild or moderate in intensity and only three AEs were severe. Only one AE was persisting at the end of the study, and most of the events that resolved did so

TABLE 20. Number and Severity of AEs by Test Treatment

Test Treatment	Number and Severity of AEs			Total
	Mild	Moderate	Severe	
THC:CBD SL drops	3	6	2	11
High CBD SL drops	1	4	1	6
High THC SL drops	5	6	0	11
Placebo	1	2	0	3
Aerosol	4	4	0	8
Nebuliser	1	5	0	6
Total	15	27	3	45

without treatment (35 events). The AEs experienced were abnormal dreams, conjunctival hyperaemia, tachycardia, pallor, sleep disorder, increased sweating, hot flushes, hyperacusis, upper abdominal pain, frequent bowel movements, increased body temperature, hunger, depressed mood, cough and hypotension.

Table 20 summarises the number and severity of AEs by test treatment.

Analysis of Adverse Events

All six subjects experienced at least two AEs each during the study. All the AEs were non-serious. Most AEs were related to the study treatment in the active groups, but more AEs were unrelated to treatment in the placebo group. The majority of AEs were mild or moderate in intensity. Only three AEs out of a total of 45 were severe and occurred when the subjects were receiving CBD:THC sublingual drops (conjunctival hyperaemia and hunger) and High CBD sublingual drops (conjunctival hyperaemia). Only one AE was persisting at the end of the study (menopausal symptoms in a 43-year-old female subject, not related to treatment), and most of the events that resolved did so without treatment.

Although the number of patients was small, there were differences between the active study treatments and placebo. Only one subject developed an AE following administration of placebo whereas three to four subjects developed AEs following administration of the active test treatments. Tachycardia, conjunctival hyperaemia and abnormal dreams were the most common AEs experienced and accounted for three, five and eight AEs, respectively, across all treatment groups. Tachycardia was the most common AE in subjects receiving High THC sublingual drops (two subjects); conjunctival hyperaemia was the most common AE in subjects receiving CBD:THC sublingual drops (two subjects); and abnormal dreams was the most common AE in subjects receiving the aerosol (two subjects) and the inhaler (two subjects). Conjunctival hyperaemia and abnormal dreams were the jointly the most common AEs in subjects receiving the nebuliser.

Abnormal dreams was the only intoxication-type AE developed by the subjects during this study. At least one subject developed them while receiving any one of the test treatments (including placebo), and four subjects developed them overall. Apart from one subject who experienced a cough during the use of the inhaler, no subjects developed any AEs that may have been related to application of the test treatment. There were no deaths, SAEs or other significant AEs during this study.

CLINICAL LABORATORY EVALUATION

Laboratory Values Over Time

The mean value of each laboratory parameter exhibited only small variations from screening to post-study. The small variations did not suggest any patterns or trends.

Individual Subject Changes

Shift tables showed no more than two shifts between categories (low, normal and high) per parameter. The small number of changes did not suggest any patterns or trends.

Individual Clinically Significant Abnormalities

There were no clinically significant abnormalities in the laboratory parameters for any subject at either screening or post-study. Subject 002's urine was positive for nitrites at screening which was considered to be a clinically relevant abnormal result. There were no other clinically relevant abnormal results.

Vital Signs, Physical Findings and Other Observations Related to Safety

The mean values of all the vital signs showed no patterns or trends and no differences from placebo. ECGs at both screening and post-study were normal for all subjects.

Conjunctival Reddening

All test treatments, including placebo but with the exception of High CBD, had a reported incidence of 17-50% of slight conjunctival reddening pre-dose. Pre-dose High CBD, no conjunctival reddening was reported. Post-dose for all test treatments including placebo, the majority of subjects (67-100%) were reported as having 'slight' or 'no' conjunctival reddening. With the exception of High CBD, moderate conjunctival reddening was reported in a maximum of two subjects (33% between 31 min and 4 h 01 min) for all test treatments including placebo. Only CBD:THC sublingual drops resulted in one subject having severe conjunctival reddening at 4 h 01 min.

Safety Conclusions

The sublingual test treatments were well tolerated by all subjects. Each of the 6 subjects experienced at least two non-serious AEs during the study, but there were no deaths, SAEs or other significant AEs. There were a total of 45 AEs, the vast majority of which were mild or moderate in intensity, only three being severe. All but one AE resolved (non-related), most (35) without treatment. Most AEs were related to the study treatment, except for subjects receiving placebo where more AEs were unrelated to treatment.

The commonest AEs were abnormal dreams, conjunctival hyperaemia and tachycardia. Abnormal dreams was the only intoxication-type AE developed by the subjects during this study and was the most common AE overall. No subjects developed any AEs that may have been related to administration of the sublingual test treatments.

The small variations in individual subject laboratory parameters and urinalyses and in the mean laboratory parameters did not suggest any patterns or trends. The mean values of all the vital signs showed no patterns or trends either and no differences from placebo. ECGs at both screening and post-study were normal for all subjects.

DISCUSSION AND OVERALL CONCLUSIONS

The sublingual test treatments were well tolerated at the doses administered by all subjects. All six subjects experienced at least two non-serious AEs during the study, but there were no deaths, SAEs or other significant AEs. All but one AE resolved without treatment. Although the number of AEs was small, subjects clearly developed more AEs when receiving the active test treatment than when receiving placebo.

No overall statistically significantly different differences were reported between each of sublingual test treatments when compared to the CBD:THC sublingual drops. However, there were few subjects in this study and due to the low concentrations of cannabinoids in plasma some PK parameters could not be calculated for some subjects.

When CBD and THC are co-administered as sublingual drops, the rate of appearance of THC is marginally increased compared to being administered as High THC suggesting that CBD may stimulate the absorption of THC. The appearance of 11-hydroxy-THC is reduced when CBD and THC are co-administered suggesting that the metabolism of THC to 11-hydroxy-THC may be reduced by CBD. THC is more extensively absorbed than CBD and no changes were seen for any sublingual drop test treatments relative to CBD:THC sublingual drops.

Administration of CBD:THC via the pressurised aerosol resulted in a slightly faster rate of absorption of CBD and THC than for the CBD:THC sublingual drops. However, overall AUCs were reduced for THC and 11-hydroxy-THC and increased for CBD.

The nebuliser resulted in a very rapid rate and relatively large extent of absorption of both CBD and THC. However it also resulted in the greatest number of adverse effects experienced by the subjects. Administration of the test treatment via the nebuliser was considered practical however, the concept of administering the test treatment via the lungs was shown to be more effective than for sublingual administration. Very low concentrations of 11-hydroxy-THC were produced following nebuliser administration indicating a reduction in metabolism of THC to 11-hydroxy-THC.

Each test treatment resulted in a reduction in subjectively assessed general well-being with the greatest effects reported following administration of CBD:THC sublingual drops. Maximum effects were experienced at approximately the same time post each dose and some effects were also reported following administration of placebo. This suggests that some of the changes in feeling of well being may be due to the excipients or a placebo effect.

The reported increase in hunger following administration of each test treatment was not unexpected as maximum hunger was reported close to lunch time. The greatest mean incidence of unpleasant effects was reported earlier than for any other effect and following administration of the nebuliser test treatment.

In conclusion, each sublingual test treatment was well tolerated by all subjects. The inhaled test treatment was not well tolerated and resulted in adverse effects.

REFERENCES

Agurell, S., M. Halldin, J. E. Lindgren, A. Ohlsson, M. Widman, H. Gillespie, and L. Hollister. 1986. Pharmacokinetics and metabolism of delta 1-tetrahydrocannabinol and other cannabinoids with emphasis on man. *Pharmacol Rev* 38 (1):21-43.

British Medical Association. 1997. *Therapeutic uses of cannabis.* Amsterdam: Harwood Academic Publishers.

Baker, P. B., T. A. Gough, and B. J. Taylor. 1983. The physical and chemical features of Cannabis plants grown in the United Kingdom of Great Britain and Northern Ireland from seeds of known origin–Part II: Second generation studies. *Bull Narc* 35 (1):51-62.

GW Pharmaceuticals. 2002. Investigators brochure for cannabis based medicine extract (CBME). Porton Down: GW Pharmaceuticals.

House of Lords. 1998. *Cannabis: The scientific and medical evidence.* London: House of Lords Select Committee on Science and Technology, Stationery Office.

Maykut, M. O. 1985. Health consequences of acute and chronic marihuana use. *Prog Neuropsychopharmacol Biol Psychiatry* 9 (3):209-38.

McPartland, J. M., and E. B. Russo. 2001. Cannabis and cannabis extracts: Greater than the sum of their parts? *J Cannabis Therapeutics* 1 (3-4):103-132.

Portenoy, R. K. 1990. Chronic opioid therapy in nonmalignant pain. *J Pain Symptom Manage* 5 (1 Suppl):S46-62.

Porter, J., and H. Jick. 1980. Addiction rare in patients treated with narcotics. *N Engl J Med* 302 (2):123.

Whittle, B. A., G. W. Guy, and P. Robson. 2001. Prospects for new cannabis-based prescription medicines. *J Cannabis Therapeutics* 1 (3-4):183-205.

A Phase I, Open Label, Four-Way Crossover Study to Compare the Pharmacokinetic Profiles of a Single Dose of 20 mg of a Cannabis Based Medicine Extract (CBME) Administered on 3 Different Areas of the Buccal Mucosa and to Investigate the Pharmacokinetics of CBME *per Oral* in Healthy Male and Female Volunteers (GWPK0112)

G. W. Guy

P. J. Robson

SUMMARY. This Phase I, open label, four-way crossover study pertains to pharmacokinetic parameters of four cannabis based medicine extracts (CBME). Sublingual, buccal and oro-pharyngeal test treatments

G. W. Guy and P. J. Robson are affiliated with GW Pharmaceuticals plc, Porton Down Science Park, Salisbury, Wiltshire, SP4 0JQ, UK.

[Haworth co-indexing entry note]: "A Phase I, Open Label, Four-Way Crossover Study to Compare the Pharmacokinetic Profiles of a Single Dose of 20 mg of a Cannabis Based Medicine Extract (CBME) Administered on 3 Different Areas of the Buccal Mucosa and to Investigate the Pharmacokinetics of CBME *per Oral* in Healthy Male and Female Volunteers (GWPK0112)." Guy, G. W., and P. J. Robson. Co-published simultaneously in *Journal of Cannabis Therapeutics* (The Haworth Integrative Healing Press, an imprint of The Haworth Press, Inc.) Vol. 3, No. 4, 2003, pp. 79-120; and: *Cannabis: From Pariah to Prescription* (ed: Ethan Russo) The Haworth Integrative Healing Press, an imprint of The Haworth Press, Inc., 2003, pp. 79-120. Single or multiple copies of this article are available for a fee from The Haworth Document Delivery Service [1-800-HAWORTH, 9:00 a.m. - 5:00 p.m. (EST). E-mail address: docdelivery@haworthpress.com].

http://www.haworthpress.com/store/product.asp?sku=J175
© 2003 by The Haworth Press, Inc. All rights reserved.
10.1300/J175v03n04_01

(GW-1000-02) consisted of 25 mg cannabidiol (CBD) + 25 mg Δ^9-tetra-hydrocannabinol (THC) per ml formulated in ethanol (eth):propylene glycol (PG) (50:50), with peppermint flavouring with a 100 μl actuation volume (total dose 10 mg CBD + 10 mg THC in 4 actuations). An oral capsule contained 2.5 mg CBD + 2.5 mg THC sprayed onto granulated lactose and encapsulated in soft gelatin capsules (total dose of 10 mg CBD + 10 mg THC 4 capsules). This study was performed in healthy volunteers in an open label, 4 period, 3-way randomised crossover followed by a non-randomised oral dose using single doses of 20 mg of CBME (10 mg CBD + 10 mg THC). In Periods 1 to 3, the test treatment was administered as a liquid spray according to the randomisation scheme (i.e., sublingually, buccally, oro-pharyngeally). In Period 4 the test treatment was delivered as an oral capsule. There was a six-day washout between each dose.

Primary objectives were to compare the pharmacokinetic profiles of cannabis based medicine extract (CBME) when administered on different areas of the buccal mucosa. Secondary objectives were to investigate the pharmacokinetic profile of CBME when administered as an oral capsule.

Concentrations of THC were higher than the corresponding levels of CBD at most time points. Concentrations of 11-hydroxy-THC exceeded the corresponding concentration of THC at most time points. By 720 min (12 h) post-dose, mean concentrations of each cannabinoid were still above the lower limit of quantification (LLOQ). There was a high degree of inter-subject and intra-subject variability in the plasma concentrations achieved.

T_{max} of CBD and THC occurred earlier following sublingual administration than oro-pharyngeal or buccal although only the difference in T_{max} of CBD compared with buccal was statistically significant. C_{max} of both CBD and THC was greatest following buccal administration although this was not statistically significant. AUC was greatest following oro-pharyngeal and was statistically significantly greater than buccal. The lower bioavailability, as measured by AUC, following buccal administration when compared to the sublingual and oro-pharyngeal routes may be related to the difficulty of spraying onto the inside of the cheek reported during the study and could be due to some loss of spray. Buccal administration of the pump action sublingual spray (PASS) test treatment resulted in a later T_{max} but greater C_{max} when compared to the sublingual and oro-pharyngeal routes. Comparison of the sublingual and oro-pharyngeal routes showed no statistically significant difference in THC or CBD pharmacokinetic parameters other than an earlier T_{max} following sublingual dosing. The oral capsule appeared to show an early T_{max} of both CBD and THC. Mean C_{max} of THC and 11-hydroxy-THC were greater, but in contrast the C_{max} of CBD was lower, than following the PASS treatments. Relative to THC, the plasma level AUC of 11-hydroxy-THC was proportionally greatest following oral capsules which could be a reflection of greater metabolism by this route. Of the PASS treatments the ratio of 11-hydroxy-THC to THC was greatest following sublingual and least following oro-pharyngeal. There was very wide inter- and to a lesser extent intra-subject variability in pharmacokinetics. Differences in mean values between the routes of administration, even when statistically significant, are small relative to the very wide range of values between subjects. The sublingual and oro-pharyngeal routes of administration appear to have the same pharmacokinetic results. The buccal pharmacokinetic parameters are lower when compared to the sublingual and oro-pharyngeal routes.

A total of 146 adverse events (AEs) occurred in 12 subjects. Two events were classified as moderate (flu-like illness and pharyngeal irritation) and the remaining 144 were classified as mild. All routes of administration were well tolerated by all subjects with no serious AEs and no withdrawals due to AEs.

The overall results indicate that administration of the liquid spray (GW-1000-02) need not be limited to sublingual administration. The oral capsule, has good bioavailability, and provided, as is the case here the formulation is not oil based, may be a viable formulation when self-titration is not necessary. *[Article copies available for a fee from The Haworth Document Delivery Service: 1-800-HAWORTH. E-mail address: <docdelivery@haworthpress.com> Website: <http://www.HaworthPress.com> © 2003 by The Haworth Press, Inc. All rights reserved.]*

KEYWORDS. Cannabinoids, cannabis, THC, cannabidiol, medical marijuana, pharmacokinetics, pharmacodynamics, multiple sclerosis, botanical extracts, alternative delivery systems, harm reduction

INTRODUCTION

Cannabis plants (*Cannabis sativa*) contain approximately 60 different cannabinoids (British Medical Association 1997), and in the UK, oral tinctures of cannabis were prescribed until cannabis was made a Schedule 1 controlled substance in the Misuse of Drugs Act in 1971. The prevalence of recreational cannabis use increased markedly in the UK after 1960, reaching a peak in the late 1970s. This resulted in a large number of individuals with a range of intractable medical disorders being exposed to the drug, and many of these discovered that cannabis could apparently relieve symptoms not alleviated by standard treatments. This was strikingly the case with certain neurological disorders, particularly multiple sclerosis (MS). The black market cannabis available to those patients is thought to have contained approximately equal amounts of the cannabinoids Δ^9-tetrahydrocannabinol (THC) and cannabidiol (CBD) (Baker, Gough, and Taylor 1983). The importance of CBD lies not only in its own inherent therapeutic profile but also in its ability to modulate some of the undesirable effects of THC through both pharmacokinetic and pharmacodynamic mechanisms (McPartland and Russo 2001). MS patients claimed beneficial effects from cannabis in many core symptoms, including pain, urinary disturbance, tremor, spasm and spasticity (British Medical Association 1997). The MS Society estimated in 1998 that up to 4% (3,400) of UK MS sufferers used cannabis medicinally (House of Lords 1998).

Cannabinoid clinical research has often focussed on synthetic analogues of THC, the principal psychoactive cannabinoid, given orally. This has not taken the possible therapeutic contribution of the other cannabinoid and non-cannabinoid plant components into account, or the slow and unpredictable absorption of cannabinoids via the gastrointestinal tract (Agurell et al. 1986). Under these conditions it has been difficult to titrate cannabinoids accurately to a therapeutic effect. Research involving plant-derived material has often reported only the THC content (Maykut 1985) of the preparations, making valid comparisons between studies difficult. GW Pharma Ltd (GW) has developed cannabis based medicine extracts (CBMEs) derived from plant cultivars that produce high and reproducible yields of specified cannabinoids. CBMEs contain a defined amount of the specified cannabinoid(s), plus the minor cannabinoids and also terpenes and flavonoids. The specified cannabinoids constitute at least 90% of the total cannabinoid content of the extracts. The minor cannabinoids and other constituents add to the overall therapeutic profile of the CBMEs and may play a role in stabilising the extract (Whittle, Guy, and Robson 2001). Early clinical studies indicated that sublingual dosing with CBME was feasible, well tolerated and convenient for titration. The concept of self-titration was readily understood by patients and worked well in practice. Dosing patterns tended to resemble those seen in the patient controlled analgesia technique used in post-operative pain control; with small doses administered as and when patients require them, up to a maximal rate and daily limit (GW Pharmaceuticals 2002). The Phase 2 experience has supported some of the wide-range of effects reported anecdotally for cannabis. It has also shown that for most patients the therapeutic benefits of CBMEs could be obtained at doses below those that cause marked intoxication (the 'high'). This is consistent with experience in patients receiving opioids for pain relief, where therapeutic use rarely leads to misuse (Porter and Jick 1980; Portenoy 1990). Onset of intoxication may be an indicator of over-titration. However the range of daily dose required is subject to a high inter-individual variability.

SATIVEX (1:1 THC:CBD CBME) was administered as an oromucosal spray, and contains an equal proportion of THC and CBD, similar to the cannabinoid profile of the cannabis thought to be most commonly available on the European black market (Baker, Gough, and Taylor 1983).

SATIVEX was administered as a liquid spray in three different areas of the mouth and 1:1 THC:CBD CBME as an oral capsule. Each formulation contained equal amounts of CBD and THC. GWPK0112 was a Phase I clinical study that aimed to investigate the relative bioavailability of CBME when administered in different areas of the oral mucosa and the absorption and bioavailability of CBME when administered orally. It was also designed to assess safety and tolerability of the test treatments.

Study Preparations

Sublingual, buccal and oro-pharyngeal test treatments (GW-1000-02) consisted of 25 mg cannabidiol (CBD) + 25 mg Δ^9-tetrahydrocannabinol (THC) per ml formulated in ethanol (eth):propylene glycol (PG) (50:50), with peppermint flavouring with a 100 µl actuation volume (total dose 10 mg CBD + 10 mg THC in 4 actuations). An oral capsule contained 2.5 mg CBD + 2.5 mg THC sprayed onto granulated lactose and encapsulated in soft gelatin capsules (total dose of 10 mg CBD + 10 mg THC 4 capsules).

Study Objectives

Primary objectives were to compare the pharmacokinetic profiles of cannabis based medicine extract (CBME) when administered on different areas of the oral mucosa. Secondary objectives were to investigate the pharmacokinetic profile of CBME when administered as an oral capsule and to assess the safety and tolerability of CBME when administered via different areas of the oral mucosa and *per oral* (po).

METHODS

The final study protocol, final Informed Consent Form, and Investigator Brochure were reviewed by PPD Development Clinic Independent Ethics Committee. Unconditional approval to conduct the study was granted on January 10, 2002. Protocol Amendment was approved by the Ethics Committee on February 6, 2002.

The planning and conduct of this study was subject to national laws and was in conformity with the current revision of the Declaration of Helsinki (October 2000, Edinburgh, Scotland), and the ICH Guidelines for Good Clinical Practice (CPMP/ICH/135/95) July 1996.

A written version of the Informed Consent Form was sent to the subjects before attending screening. At the screening visit and prior to any screening procedures being carried out, the Informed Consent Form was presented verbally to the subjects. The Informed Consent Form detailed no less than: the exact nature of the study; the implications and constraints of the protocol; the known side effects that they might expect and any risks involved in taking part; subjects were advised that they would be free to withdraw from the study at any time for any reason without prejudice to future care. Subjects were allowed sufficient time and the opportunity to question the Principal Investigator, their General Practitioner or other independent parties to decide whether they wanted to participate in the study. Written Informed Consent was then obtained by means of subject signature, signature of the person who presented Informed Consent and, if different, the Principal Investigator. A copy of the signed Informed Consent Form was given to the subject and the original signed form is retained in the study site files.

This study was conducted at PPD Development Clinic, 72 Hospital Close, Evington, Leicester, LE5 4WW. The plasma concentration analysis was carried out at ABS Laboratories Ltd, Wardalls Grove, Avonley Road, London, SE14 5ER. The Sponsor for this study was GW Pharma Ltd, Alexander House, Forehill, Ely, Cambridgeshire CB7 4ZA. The test treatments used in this study were formulated by G-Pharm Ltd.

Overall Study Design and Plan–Description

The study was an open label, 4 period, 3-way randomised crossover followed by a non-randomised *po* dose using single doses of 20 mg of CBME (10 mg CBD + 10 mg THC). In Periods 1 to 3 the test treatment was administered to subjects as a liquid spray according to the pre-determined randomisation scheme sublingually (Treatment A), buccally

(Treatment B: inside of cheek), oro-pharyngeally (Treatment C: sprayed generally in mouth), and in Period 4, as an oral capsule (Treatment D). Treatments A, B and C were administered as four actuations (sprays) each five minutes apart. The oral capsule was administered *po* as four capsules each five minutes apart. There was a minimum of six days washout between each. The liquid sprays (GW-1000-02) were formulated in 50% ethanol:50% propylene glycol (PG) at a concentration of 25 mg CBD + 25 mg THC/ml, with peppermint flavouring. The 1:1 THC:CBD capsules were produced as 2.5 mg CBD + 2.5 mg THC sprayed onto granulated lactose in soft gelatin capsules.

Twelve healthy subjects (six male and six female) who complied with all the inclusion and exclusion criteria were required to complete the study in its entirety.

Discussion of Study Design

The present route of administration of CBME used to date in patient studies has been limited to sublingual sprays. Due to the limitation of using a small area of the oral cavity there is at least a potential for mucosal tenderness, lesions or other adverse reactions when used chronically. Therefore the different oral mucosal routes of administration were chosen to assess the plasma concentration-time profiles and pharmacokinetic parameters in relation to the sublingual route.

The oral capsule was chosen to make a preliminary assessment of the plasma concentration-time profiles and pharmacokinetic parameters following oral administration. The dose of CBME administered in this study (10 mg CBD + 10 mg THC) was chosen as this is representative of the dosage of the test treatment when used by patients in a self-titrated regime. It is also known to be well tolerated by subjects and produce quantifiable concentrations of cannabinoids in plasma.

GW specified that only subjects with previous experience with the effects of cannabis be included in their Phase I trials to ensure that subjects recognise the effects they may experience as a result of the CBME given. A crossover design was chosen to enable both inter- and intra-subject comparisons of pharmacokinetic data. The study design was open label as blinding was not possible with different routes of administration. A six-day washout ensured all cannabinoids were below the limit of quantification and assisted in the scheduling of the study in the clinical unit.

Inclusion Criteria

For inclusion in the study subjects were required to fulfil all of the following criteria to ensure they were normal healthy subjects and agreed to participate as per the protocol:

  i.   Healthy and aged between 18 and 50 years
  ii. Had a body mass index (BMI) between 19 and 30 kg/m²
  ii. Had given written informed consent
  iv. Had experienced the effects of cannabis more than once
  v.   Agreed to comply with all the study requirements and restrictions
  vi. Agreed to use barrier methods of contraception throughout the study and for 3 months post-dose

Subject demographics and habits are noted in Table 1 and 2.

Exclusion Criteria

To ensure they were normal and healthy, subjects were deemed not acceptable for participation in the study if any of the following criteria applied:

  i.   Had any cardiovascular, haematological, hepatic, gastro-intestinal, renal, pulmonary, neurological or psychiatric disease which in the opinion of the Investigator was significant
  ii.   Had a history or presence of schizophrenic-type illness
  iii. Had a history of drug or alcohol abuse in the past 12 months
  iv.   Had a history of allergy to cannabis and/or its metabolites
  v.   Had used cannabis in any form in the 30 days prior to dosing
  vi.   Had an abnormal blood or urinalysis result at screening which in the opinion of the Investigator was clinically significant
  vii. Had a positive drug screen result (including cannabis) at screening
  viii. Had a resting blood pressure > 150/95 or < 90/50 mmHg and a pulse < 40 or > 120 b.p.m.
  ix.   Had taken a course of prescribed medication (with the exception of oral or depot contraceptives) in the 4 weeks prior to dosing
  x.   Had taken any over-the-counter or prescription medication (with the exception of oral or depot contraceptives) in the 14 days prior to dosing. If currently taking vitamins or paracetamol subjects were asked to discontinue use at screening

   xi. Had been hospitalised in the 3 months prior to dosing

   xii. Had lost or donated > 400 ml of blood in the 3 months prior to dosing

  xiii. Smoked ≥ 5 cigarettes or used ≥ 1/4 ounces of tobacco per day

  xiv. Had participated in a clinical trial in the 3 months prior to dosing

   xv. Regularly consumed = 28 (males) or = 21 (females) units of alcohol per week

  xvi. Was pregnant or lactating at the time of screening

 xvii. Planned to become pregnant during or for three months after completion of the study

Study Restrictions

Subjects were required to abstain from the following for the duration of the study:

   i. All foods and beverages containing caffeine and alcohol for 24h pre-each dose until the end of each confinement period

  ii. Taking any drugs, including drugs of abuse, prescribed and/or over-the-counter medications for the duration of the study

 iii. Smoking/using cigarettes/tobacco products during each confinement period

 iv. Donating blood or participating in another clinical study in the 3 months after completion of the study

Removal of Subjects from Therapy or Assessment

The subjects were free to withdraw from the study without explanation at any time and without prejudice to future medical care. Subjects may have been withdrawn from the study at any time if it was considered to be in the best interest of the subject's safety.

TABLE 1. Demographic Data

Statistic	Age (years)	Height (m)	Weight (kg)	BMI (kg/m^2)
Mean	36.5	1.721	72.38	24.33
Median	36.5	1.73	71.55	24.3
SD	8.38	0.0902	10.785	1.80
Minimum	21	1.58	57.9	21.8
Maximum	48	1.89	98.3	27.5

TABLE 2. Demographics and Habits

Variable	Frequency
Sex:	
Male	6
Female	6
Race:	
Caucasian	11
Mixed Race	1
Smoking:	
None	6
≥ 5 cigarettes/day	6
Alcohol:	
None	0
< 14 units/week	10
< 21 units/week	2
Previous cannabis use: Effects experienced more than once	
Yes	12
No	0

Variable	Frequency
Drugs of Abuse:	
Negative	12
Positive	0
Pregnancy Test:	
Negative	6
Not Required	0
Contraception:	
Yes	12
No	0
CS Blood/Urine Result:	
Yes	0
No	12

CS = clinically significant

TEST TREATMENTS

Treatments Administered

A total single dose of 10 mg CBD + 10 mg THC was administered sublingually, buccally, oro-pharyngeally or *po* to each of 12 subjects on four occasions. Each single dose (10 mg CBD + 10 mg THC) consisted of a series of four actuations of 100 μl (2.5 mg CBD + 2.5 mg THC per actuation) or four capsules (2.5 mg CBD + 2.5 mg THC per capsule) and each actuation/capsule was administered five minutes apart. Every subject received each of the test treatments once. Each vial and capsule blister pack was labelled with no less than subject number, period number, unit number and expiry date.

For the sublingual, buccal and oro-pharyngeal test treatment (Periods 1-3) subjects were randomised to a dose sequence using a Williams Square Design provided by GW. All subjects received the oral capsule in Period 4. All subjects received a single dose of one test treatment in each period.

Selection of Doses in the Study

The dose given has been previously used in GW studies and has been shown to be both well tolerated and produce quantifiable plasma drug concentrations. The dosing regime was chosen as it has been well tolerated by subjects and in general is a reflection of the dosing regimen used in patient studies when the patients are self-titrating.

Selection and Timing of Dose for Each Subject

The test treatments were administered in the morning of each dosing day according to the randomisation scheme. Subjects were dosed in the morning to allow blood samples to be taken and procedures to be carried out up to 12 h post-dose without confining the subjects to the clinical unit overnight. A minimum of six days washout between each dose was specified, as previous data and drug of abuse screens have indicated that concentrations of each cannabinoid from a single dose of CBME are below the limit of quantification by this time. The study was open label.

Subjects were required to abstain from taking any medication, over the counter and prescribed for 14 and 28 days, respectively, prior to dosing until completion of the study unless recommended by their General Practitioner. If any subject took concomitant medications during the restriction period it was noted in the CRF and Investigator judgement as to the subjects continued eligibility was made.

Test Treatment Compliance

Subjects were dosed by the Principal Investigator or suitably trained designee. For the sublingual, buccal and oro-pharyngeal routes subjects were instructed to allow each actuation to absorb and not to swallow if possible. For the *po* route, each capsule was placed on the subject's tongue and they were instructed to swallow the capsule using the glass of water (50 ml) provided to wash each capsule down. Following administration of each capsule the person administering the dose checked the subject's mouth to ensure the capsule had been swallowed. The actual time of administration of each actuation/capsule was recorded in the CRF and the dosing procedure was witnessed by a dose verifier. All subjects received all of the scheduled doses and there were no deviations from dosing target times.

STUDY PROCEDURES

Pre-Study Screening

Subjects were required to undergo a pre-study screen no more than 21 days prior to first dose administration to determine their eligibility to take part in the study. Only those subjects who were healthy and were willing to comply with all the study requirements were deemed eligible for participation. The screening procedures comprised the assessments/ measurements shown below.

Demographic Data

The subjects' date of birth, age, sex, race, height, weight, body mass index (BMI), previous cannabis experience, tobacco and alcohol consumption were recorded (Tables 1-2).

Concomitant Medications and Medical History

Subjects were asked to provide details of any drugs, vitamins or medications they had taken in the four weeks prior to screening or were taking at the time of screening. Details of their previous medical history were also recorded. Subjects underwent a physical examination to determine if there were any abnormalities in any body systems. Blood pressure (systolic/diastolic) and pulse were measured after the subject had been seated for no less than two minutes. Oral temperature was also measured. A 12-lead ECG (electrocardiograph) was taken for each subject. At least the following ECG parameters were recorded: HR (heart rate), PR, QT_C and QRS intervals.

Subjects were required to provide a urine sample for routine urinalysis including protein, glucose, ketones, bilirubin, nitrites, blood, urobilinogen, haemoglobin and pH. Microscopy was required to be carried out on any abnormal samples. A pregnancy test was carried out using an HCG Pregnancy Test on all urine samples from female subjects. The samples provided (male and female) were also screened for the drugs of abuse including methadone, benzodiazepines, cocaine, amphetamine, THC, opiates, and barbiturates.

A 4.7 ml blood sample was taken in an ethylenediaminetetraacetic acid (EDTA) blood tube for haematology analysis. A 2.7 ml blood sample was taken in a gel blood tube for routine clinical chemistry analysis.

A blood sample (2.7 ml) was taken in a gel blood tube to screen for the presence of Hepatitis B and/or C.

Pre-Dose Procedures

Subjects were required to arrive at the clinic approximately one hour prior to dosing for each study period. Each subject's health status was updated and pre-dose procedures (health status update, blood pressure and pulse, alcohol and drug of abuse screen and pregnancy test for female subjects) were carried out. Only subjects who complied with the requirements of the study were accepted for inclusion in the study.

Blood Sampling for Plasma CBME Concentration Analysis

Blood samples (4.5 ml) were collected into lithium heparin blood tubes via indwelling cannula or individual venipuncture. Samples were placed immediately into an ice bath until centrifuged (3000 rpm for 10 min at 4°C). The resultant plasma was decanted into two identical pre-labelled silanised amber glass plasma tubes and placed in a freezer at -20°C. Blood samples were collected pre-dose and at 15, 30 and 45 min, 1, 1.25, 1.5, 1.75, 2, 2.25, 2.5, 2.75, 3, 3.5, 4, 4.5, 5, 5.5, 6, 8 and 12 h post start of dose.

Plasma concentrations of CBD, THC and 11-hydroxy-THC were measured in each plasma sample. Urine samples were collected in individual 1 L polypropylene containers. Samples were placed in a refrigerator at $+4$°C (range of 0 to 10°C) until the end of each collection period. Samples were then pooled by collection period and the total volume recorded. Sub-samples (2 $\times$ 20 ml) were retained (stored frozen at -20°C) for analysis and the remainder of the urine discarded. Urine samples were collected for the following time periods: -1 to 0, 0 to 0.5, 0.5 to 1, 1 to 3, 3 to 6 and 6 to 12 h post-dose. Urine concentrations of 11-COOH THC were measured in each urine sample

Safety Assessments

Each subject was required to provide a urine sample for a urine drug screen at check in for each dosing period. The drug screen was required to be negative for all drugs pre-dose Period 1. For Periods 2 to 4, positive THC results may have occurred due to administration of test treatment in the previous period and therefore screening for THC was not carried out. The urine sample was required to be negative for all other drugs tested for the subject to be eligible to continue.

Subjects' blood pressure and pulse were measured pre-dose and at 30 min then 1, 1.5, 2, 2.5, 3, 3.5, 4, 4.5, 5, 6, 8 and 12 h post start of dosing. A 12-Lead ECG was taken for each subject at the following times: pre-dose and at 30 min, 1, 1.5, 2, 2.5, 3, 3.5, 4, 4.5, 5, 6, 8 and 12 h post-dose.

Adverse Events

Subjects' health was monitored continuously throughout the study for Adverse Events (AEs). All AEs were recorded in the CRF. In addition, subjects' health was monitored by asking non-leading questions pre-dose and at the following times post-dose: 15, 30, 45 min, 1, 1.25, 1.5, 2, 2.5, 3, 4, 5, 6, 8 and 12 h post-dose. All AEs were noted and followed to resolution or at the discretion of the Investigator.

Pregnancy Test

A pregnancy test was carried out using an HCG Pregnancy Test for all female subjects on the urine samples provided at check-in for each study period. The test was required to be negative for the subject to continue in the study.

Palatability/Dose Questionnaire

As soon as possible after the dosing was completed, subjects were asked to complete a questionnaire about the palatability and physical sensation of the test treatment experienced during and immediately after dosing.

Food and Beverages

On study dosing days, subjects were required to abstain from consuming food and beverages for 15 min before the first actuation and 15 min post last actuation (Periods 1-3 only). For Period 4 (capsule dosing) the subjects were not allowed to consume food and beverages for 15 min before dosing and were only allowed to drink the 4 × 50 ml glasses of water provided for dosing until 15 min after dosing was completed.

Lunch and dinner were provided for the subjects at approximately 4 h and 10 h post-dose, respectively. Snacks, e.g., digestive biscuits, were provided *ad libitum* throughout each confinement period as required.

Subjects were required to drink 100 ml of tap water hourly (with the exception of the food and beverage restriction period) from 1 h pre-dose to 10 h post-dose. Decaffeinated beverages were provided *ad libitum* throughout each confinement period as required.

Check-Out Procedures

After completion of the 12 h study procedures at the end of Periods 1, 2 and 3, and if deemed well enough to leave, subjects were discharged from the clinical unit. Prior to discharge, ongoing AEs were updated and follow up arranged if required. Prior to Period 4 discharge, subjects were required to undergo a physical examination, blood samples were taken for haematology and clinical chemistry analyses and a urine sample taken for urinalysis. In addition a 12-lead ECG was taken and vital signs recorded as per screening. Ongoing AEs were updated and if required arrangements were made to follow up with the subjects after they left the clinical unit.

DATA QUALITY ASSURANCE

Study Monitoring

All details regarding the study were documented within individual Case Report Forms (CRFs) provided by GW for each subject. All data recorded during the study were checked against source data and for compliance with GCP (Good Clinical Practice), internal SOPs (Standard Operating Procedures), working practices and protocol requirements. Monitoring of the study progress and conduct was ongoing throughout the study. Monitoring was conducted by GW Clinical Department staff and was conducted according to GW SOPs. Haematology and clinical chemistry analyses were carried out by Leicester General Hospital.

Investigator Responsibilities

The Investigator was responsible for monitoring the study conduct to ensure that the rights of the subject were protected, the reported study data was accurate, complete and verifiable and that the conduct of the study was in compliance with ICH GCP.

At the end of the study the Principal Investigator reviewed and signed each CRF declaring the data to be true and accurate. If corrections were made after review the Investigator acknowledged the changes by re-signing and dating the CRF.

Clinical Data Management

Data were double entered into approved data tables in Microsoft® Excel 2000 software. Manual checks for missing data and inconsistencies were carried out according to GW's document Data Handling Manual: Manual Checks and queries were raised for any resulting issues. Once the data were clean, i.e., no outstanding queries, then QC checks of 100% of the data for a 10% sample of the patients were conducted in order to make a decision on the acceptability of the data. Any errors were resolved and any error trends across all patients were also corrected. Clinical Quality Audits were carried out.

STATISTICAL METHODS PLANNED IN THE PROTOCOL AND DETERMINATION OF SAMPLE SIZE

Statistical and Analytical Plans

With the exception of a SAP being produced prior to carrying out statistical analyses, the statistical analyses were carried out in accordance with the protocol.

Significance Testing and Estimation

The primary analysis was estimation of the pharmacokinetic parameters and thus 95% confidence intervals (CI), in line with current guidelines, are provided for each contrast. Hypothesis testing was secondary in this study. All tests were two-sided.

Pharmacokinetic Analysis

No more than one blood sample per period was omitted for any subject therefore all subjects were considered to be evaluable for pharmacokinetic analysis and were included in the final dataset. All analyses and summary statistics were carried out and derived using SAS v8. A summary of the mean plasma concentration data is contained in Table 3. Mean pharmacokinetic parameters are contained in Table 4.

Individual plasma concentration-time data and mean profile (mean and standard deviation (SD)) for THC, 11-hydroxy-THC and CBD for each subject were recorded. Plasma concentration-time data were summarised by test treatment group at each time point. Descriptive statistics (number (N), mean, SD, geometric mean, minimum and maximum) were formulated by test treatment for the raw values. Descriptive statistics were calculated for the raw values (N, arithmetic mean, SD, co-efficient of variation (CV%)) and also for the log transformed data (geometric mean, mean of logs and SD of logs).

The pharmacokinetic parameters area under the curve from zero to

TABLE 3. Mean Plasma Concentration Data

Time (min)	CBD				THC				11-Hydroxy THC			
	SL	Buccal	o.p.	p.o.	SL	Buccal	o.p.	p.o.	SL	Buccal	o.p.	p.o.
0	0.00	0.00	0.00	0.00	0.00	0.00	0.01	0.00	0.00	0.00	0.00	0.00
15	0.06	0.04	0.00	0.06	0.05	0.04	0.01	0.08	0.03	0.02	0.00	0.04
30	0.82	0.26	0.24	1.13	1.17	0.47	0.42	2.94	1.12	0.46	0.38	2.59
45	1.00	0.54	1.18	1.61	1.97	1.21	2.68	4.97	2.71	1.60	1.77	5.82
60	1.30	1.18	1.35	1.44	2.83	2.52	3.20	4.29	4.01	2.71	2.92	6.19
75	1.55	1.20	1.80	1.64	3.41	2.74	4.17	4.23	4.93	3.25	4.02	6.75
90	1.60	1.01	1.76	1.61	3.42	2.47	3.98	3.94	5.34	3.43	4.78	6.50
105	1.73	0.99	1.73	1.41	3.56	2.45	3.71	3.09	5.32	3.78	4.65	5.78
120	1.79	1.03	1.56	1.20	3.92	2.69	3.39	2.57	5.35	3.88	4.36	5.13
135	1.53	1.04	1.57	1.25	3.32	2.57	3.30	2.34	4.71	3.70	4.27	4.71
150	1.36	1.06	1.39	1.07	2.87	2.64	2.96	2.04	4.65	4.00	4.01	4.18
165	1.26	1.08	1.34	1.09	2.48	2.63	2.78	2.02	4.56	4.15	3.90	3.71
180	1.23	1.01	1.31	0.97	2.59	2.34	2.69	1.80	4.55	4.05	3.90	3.59
210	0.96	0.96	0.96	0.66	1.80	2.00	1.98	1.17	3.81	3.37	3.20	2.69
240	0.72	1.34	0.78	0.52	1.27	2.36	1.79	0.88	3.03	3.23	2.97	2.30
270	0.67	1.28	1.02	0.57	1.47	2.04	2.31	0.79	2.81	3.10	3.54	1.91
300	0.55	0.73	0.93	0.35	1.15	1.17	2.01	0.56	2.38	2.32	3.11	1.54
330	0.38	0.50	0.71	0.25	0.79	0.82	1.41	0.39	1.76	1.82	2.40	1.23
360	0.33	0.37	0.51	0.21	0.72	0.64	1.02	0.31	1.62	1.45	2.02	1.08
480	0.22	0.22	0.26	0.13	0.33	0.31	0.40	0.17	0.99	0.88	1.06	0.73
720	0.11	0.11	0.15	0.12	0.13	0.12	0.14	0.13	0.56	0.47	0.56	0.48

SL = sublingual   o.p. = oro-pharyngeal   p.o. = per oral
NB. Oral capsule administered in Period 4 (except Subject 010)

TABLE 4. Mean Pharmacokinetic Parameters

Treatment	T_{max} (min)	C_{max} (ng/ml)	$t_{1/2}$ (min)	$AUC_{0\text{-}t}$ (ng/ml.min)	$AUC_{0\text{-}\infty}$ (ng/ml.min)
Mean Pharmacokinetic Parameters for CBD					
Sublingual	98	2.50	86.35	408.53	427.33
Buccal	168	3.02	108.39	384.13	407.79
Oro-Pharyngeal	123	2.61	105.50	469.08	496.98
per oral	76	2.47	65.41	345.68	362.04
Mean Pharmacokinetic Parameters for THC					
Sublingual	98	5.54	105.70	808.78	837.25
Buccal	144	6.14	80.47	751.23	770.62
Oro-Pharyngeal	134	6.11	81.20	962.68	985.12
per oral	63	6.35	71.71	705.38	724.79
Mean Pharmacokinetic Parameters for 11-Hydroxy-THC					
Sublingual	95	6.24	128.84	1522.09	1632.46
Buccal	144	6.13	114.34	1293.14	1362.12
Oro-Pharyngeal	144	6.45	125.78	1477.82	1580.33
per oral	81	7.87	100.10	1410.99	1480.39

NB. Oral capsule administered in Period 4 (except Subject 010)

infinity ($AUC_{0\text{-}\infty}$), area under the curve from zero to t ($AUC_{0\text{-}t}$) and maximum concentration (C_{max}) were log transformed prior to analysis and analysed using the first three periods only. The analysis of variance (ANOVA) model included terms for subject, period and treatment. Least squares means for the treatments were transformed back to the original scale and presented as geometric means. The differences for each of the three pairwise contrasts were exponentiated to express them as ratios of geometric means with 95% confidence intervals.

Time to maximum concentration (T_{max}) and half-life ($t_{1/2}$) were analysed and transformed using the same model as above. The elimination rate constant (K_{el}) is presented descriptively only. Oral capsule data are presented descriptively.

No statistical comparisons were carried out on the urine data.

SAFETY ANALYSIS

Adverse Events

All Adverse Events were coded by Medical Dictionary of Regulatory Activities (MedDRA) and presented by System Organ Class (SOC) and

Preferred Term (PT). For each table, the distribution (n and %) of subjects are presented. The following summary tables were produced: overview summary of treatment-related Adverse Events and all causality Adverse Events.

Clinical Laboratory Tests

For each of the haematology and clinical chemistry parameters, descriptive statistics (N, mean, SD, median, minimum and maximum) were calculated and summarised by treatment group at screening and post-study. In addition, descriptive statistics were calculated and summarised for the change from screening.

Listings of clinical chemistry parameters at screening and post-study are presented in Table 5. Abnormal values were designated as H (high) or L (low) in the individual data listings based on the Normal Labora-

TABLE 5. Mean Clinical Chemistry Data

Variable	Mean pre-study (SD) n = 12	Mean post-study (SD) n = 12	Difference (SD) n = 12
AST (iu/l)	20.4 (4.60)	16.2 (3.41)	−4.3 (3.14)
ALT (iu/l)	17.4 (7.91)	15.2 (7.17)	−2.3 (2.34)
Alk phosph. (iu/l)	66.4 (14.64)	61.7 (20.11)	−4.8 (15.26)
GGT (iu/l)	19.2 (6.71)	14.5 (4.70)	−4.7 (3.55)
Total Bilirubin (µmol/l)	11.4 (5.52)	6.0 (2.86)	−5.4 (4.19)
Albumin (g/l)	44.7 (2.77)	38.9 (2.27)	−5.8 (3.93)
Total Protein (g/l)	71.0 (5.06)	63.9 (3.92)	−7.1 (5.12)
Urea (mmol/l)	4.78 (1.011)	4.55 (0.922)	−0.23 (0.916)
Creatinine (µmol/l)	79.3 (10.01)	87.8 (10.08)	8.4 (7.63)
Adjusted Calcium (mmol/l)	2.247 (0.0785)	2.351 (0.1435)	0.104 (0.0914)
Sodium (mmol/l)	137.8 (1.19)	138.3 (1.22)	0.4 (1.24)
Potassium (mmol/l)	4.08 (0.299)	4.11 (0.178)	0.03 (0.281)

tory Reference Ranges. Shift tables were constructed to determine the categorical shifts from screening to post-study. For vital signs and/or blood pressure and pulse descriptive statistics (N, mean, SD, median, minimum and maximum) were calculated and summarised at each time point by treatment group. In addition, the calculations were performed for the absolute change from pre-dose.

For each of the ECG parameters (heart rate, PR interval, QT_c interval and QRS width), descriptive statistics (N, mean, SD, median, minimum and maximum) were calculated and summarised at each time point by treatment group. In addition, the calculations were performed for the absolute change from pre-dose.

No concomitant medications were taken by any subjects throughout the study. No formal sample size calculation was carried out for this study. Only one minor change to the planned analyses occurred; the planned CI for statistical analyses (90%) was changed to 95%.

Study Subjects

Six healthy male and six healthy female subjects were required to complete the study in its entirety (see demographic data). Six male and six female subjects were randomised and all of those subjects completed the study. No subjects withdrew from the study and no replacements were required.

Protocol Deviations

The following protocol deviations which occurred during the study required investigator judgement:

1. A 4.7 ml blood sample was taken in a gel blood tube blood from each subject at pre-study screening to screen for the presence of Hepatitis B and/or C. The blood sampling for this analysis and results were retained with the individual subject CRFs.
2. Subject 010 was ill for dosing Period 3, however, did wish to continue in the study and a decision was made to delay the subject by one week. The dose to be received in Period 4 would have expired prior to the dosing date therefore the doses for Period 3 and 4 were reversed so that the subject received the oral capsule in Period 3. The actual dates of dosing for each period were recorded in the CRF.

3. A SAP was not produced prior to the statistical analyses being carried out. Statistical analyses were carried as detailed in this report.

The protocol deviations noted are not considered to affect the integrity of the study.

Plasma and Urine Concentration and Plasma Pharmacokinetic Evaluation

All twelve subjects (001 to 012) who were randomised in the study were included in the data analysis. All subjects included in the study complied with all demographic and baseline requirements for inclusion.

Measurements of Compliance

Each test treatment was administered by suitably trained study site clinical staff. No deviations to the dosing regimen were noted for any subject through out the study. The site clinical staff reported a slight difficulty in aiming the buccal dose onto the inside of the cheek, however, each dose was administered with no deviations.

INDIVIDUAL PLASMA CONCENTRATION DATA AND PHARMACOKINETIC RESULTS

Analysis of Plasma Concentration Data

Plasma samples were analysed for CBD, THC and 11-hydroxy-THC according to the analytical protocol. Plasma concentration results were produced in tabular form and concentration-time graphs were produced from these data. The LLOQ for this study was 0.1 ng/ml. Data below the LLOQ are presented as < 0.1 and the actual value measured is presented in parentheses. The actual values measured were used when creating graphs.

The mean values listed in Table 3 show that CBD, THC and 11-hydroxy-THC were all detectable in plasma at around 15-30 min after dosing. Plasma concentrations generally increased to a peak between 45 and 120 min, although following buccal dosing the mean peak of CBD was later, and thereafter diminished though low concentrations were still detectable 720 min after dosing.

Plasma levels of THC (Figure 1) exceeded the corresponding level of

CBD (Figure 2) at almost all time points by a factor of approximately 2 except early and late in the sampling schedule when concentrations of both were low. Approximately 60 min after dosing plasma levels of THC were exceeded by the levels of 11-hydroxy-THC, its principal metabolite, except following oro-pharyngeal dosing when this was delayed and did not occur until after 90 min (Figure 3).

The SDs for the mean plasma concentrations of each cannabinoid, indicate a relatively high inter-subject variability in the rate and extent of absorption (Figures 4-16). This inter-subject variation in the extent of absorption does not seem to be consistently predictable from one treatment to another due to additional intra-subject variability.

Analysis of Urine Concentration Data

Urine samples were analysed for 11-COOH THC according to the analytical protocol. Mean urine concentrations are listed in Table 6 and summarised graphically in Figure 17. The LLOQ (lower limit of quantification) for this study was 0.5 ng/ml. Data below the LLOQ are presented as < 0.5 and the actual value measured is presented in parentheses. Urine samples were collected in polypropylene containers and the binding of cannabinoids to this material is unknown. Therefore the

FIGURE 1. GWPK0112 Mean CBME THC PK Data Following Administration of CBME via Different Routes

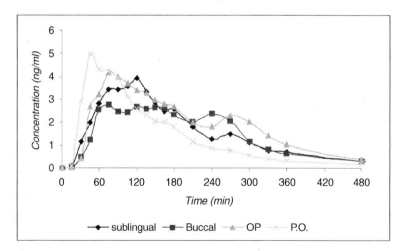

FIGURE 2. GWPK0112 Mean CBME CBD PK Data Following Administration of CBME via Different Routes

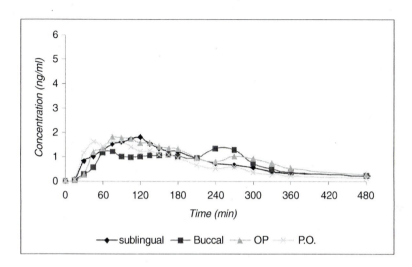

FIGURE 3. GWPK0112 Mean CBME 11-Hydroxy-THC PK Data Following Administration of CBME via Different Routes

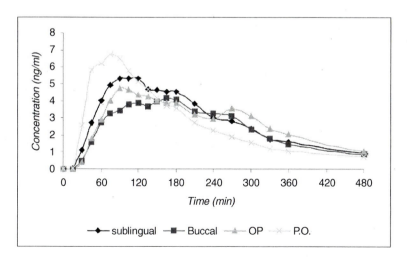

FIGURE 4. GWPK0112 Mean CBME THC PK Data Following Sublingual Administration

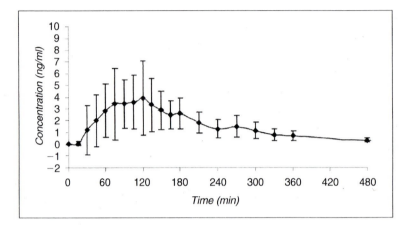

FIGURE 5. GWPK0112 Mean CBME THC PK Data Following Buccal Administration

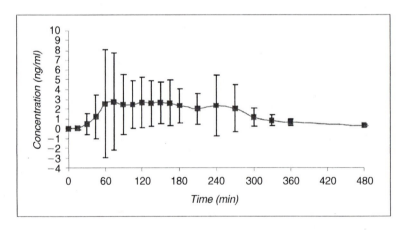

reliability of the data is not known. Pre-dose, some subjects had quantifiable amounts of 11-COOH THC in urine. Mean pre-dose concentrations were: 0.21, 0.27, 0.36 and 0.91 ng/ml in the urine samples collected in the hour prior to sublingual, buccal, oro-pharyngeal or oral capsule dosing, respectively.

FIGURE 6. GWPK0112 Mean CBME THC PK Data Following Oro-Pharyngeal Administration

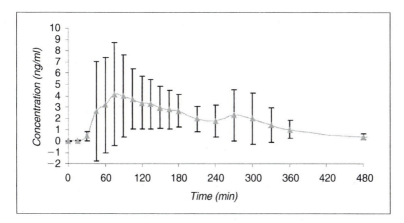

FIGURE 7. GWPK0112 Mean CBME THC PK Data Following P.O. Administration (Capsule)

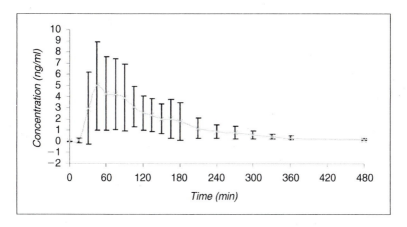

No unchanged CBD or THC were detected in urine following administration of each test treatment. A metabolite of THC (11-COOH THC) was detected and was quantified. Following each treatment the excretion of 11-COOH THC was low up to one hour after dosing, increased markedly during the 1-3 h post-dose period and increased further during

FIGURE 8. GWPK0112 Mean CBME Plasma Concentration Results Following Sublingual Administration (Subjects 1-12)

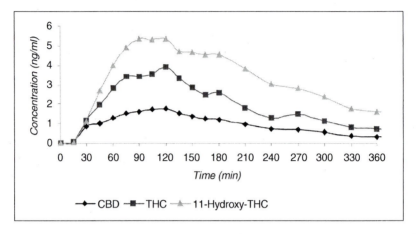

FIGURE 9. GWPK0112 Mean CBME CBD PK Data Following Sublingual Administration

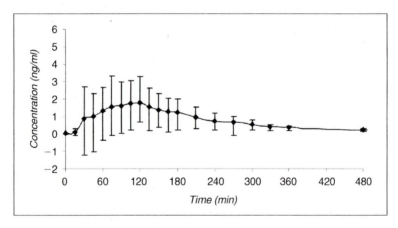

the 3-6 h period before declining again during the 6-12 h post-dose period (Table 6). All four test treatments showed a similar pattern. The highest total mean excretion apparently was achieved following administration of the oral capsule followed by the sublingual spray, buccal spray and finally oro-pharyngeal spray. However, as excretion of 11-hydroxy-THC was apparently not complete after the 6-12 h collection period, these findings should be interpreted with caution.

FIGURE 10. GWPK0112 Mean CBME CBD PK Data Following Buccal Administration

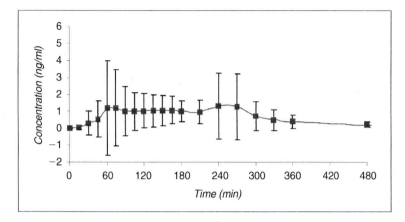

FIGURE 11. GWPK0112 Mean CBME CBD PK Data Following Oro-Pharyngeal Administration

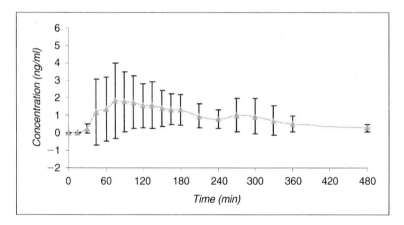

Analysis of Pharmacokinetic Parameters

Pharmacokinetic parameters were calculated using WinNonlin® Professional 3.1. The model used was a non-compartmental, linear trapezoidal analysis. Values below the LLOQ are not considered reliable and therefore were not used when calculating PK parameters. Mean pharmacokinetic values are presented in Table 4 and displayed in the graphs.

FIGURE 12. GWPK0112 Mean CBME CBD PK Data Following P.O. Administration (Capsule)

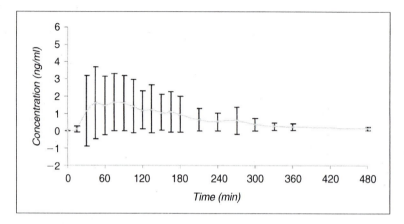

FIGURE 13. GWPK0112 Mean CBME 11-Hydroxy THC PK Data Following Sublingual Administration

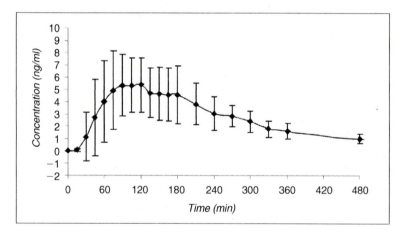

Analysis of PASS Sublingual, Buccal and Oro-Pharyngeal Pharmacokinetic Parameters

Mean T_{max} of both THC (Table 7) and CBD (Table 8) occurred earlier following sublingual administration (98 min) than oro-pharyngeal (123 min CBD, 134 min THC) (Tables 9 and 10) or buccal (168 min

FIGURE 14. GWPK0112 Mean CBME 11-Hydroxy-THC PK Data Following Buccal Administration

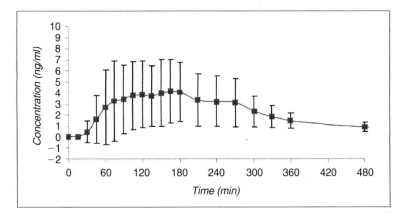

FIGURE 15. GWPK0112 Mean CBME 11-Hydroxy-THC PK Data Following Oro-Pharyngeal Administration

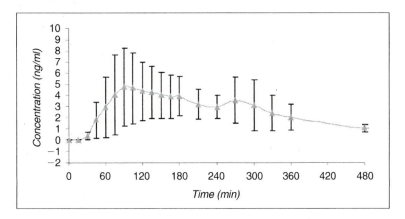

CBD, 144 min THC) (Tables 11 and 12) though only the difference in CBD T_{max} between buccal and sublingual administration reached statistical significance (p = 0.0059). C_{max} of both THC and CBD was greatest following buccal administration then oro-pharyngeal and finally sublingual, although none of the differences reached statistical significance. AUC_{0-t} and $AUC_{0-\infty}$ of both THC and CBD were greatest following oro-pharyngeal administration followed by sublingual then

FIGURE 16. GWPK0112 Mean CBME 11-Hydroxy-THC PK Data Following P.O. Administration (Capsule)

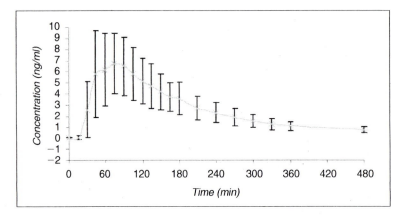

TABLE 6. Mean Excretion of 11-COOH THC in Urine (ng/ml) per Time Period

Time Period (h)	Test Treatment			
	Sublingual	Buccal	Oro-Pharyngeal	Oral Capsule
−1-0	0.21	0.27	0.36	0.91
0-0.5	0.05	0.20	0.32	0.29
0.5-1	1.10	0.75	0.89	1.61
1-3	38.45	13.55	34.53	44.17
3-6	139.33	99.29	74.40	168.84
6-12	104.08	75.97	73.30	87.12
Total	283.22	190.03	183.8	302.94

buccal dosing. The differences in AUC_{0-t} and $AUC_{0-\infty}$ of THC between oro-pharyngeal and buccal dosing were statistically significant (AUC_{0-t} p = 0.0024 and $AUC_{0-\infty}$ p = 0.0018). The bioavailability of THC was approximately twice that of CBD irrespective of the site of application.

There were significant differences in the pharmacokinetic parameters of 11-hydroxy-THC between the different administrations. T_{max} of 11-hydroxy-THC occurred statistically significantly earlier (95 min) after sublingual dosing (Table 13) than buccal (144 min, p = 0.038) (Table 14) or oro-pharyngeal (144 min, p = 0.038) (Table 15). There were no statistically significant differences in C_{max} of 11-hydroxy-THC between treatments. AUC_{0-t} and $AUC_{0-\infty}$ were significantly lower after

FIGURE 17. Mean Urine 11-COOH THC Concentrations (ng/ml) in Urine Following Administration of Each Test Treatment

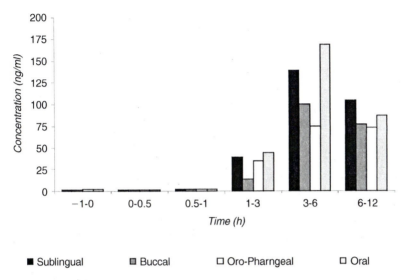

Mean (Subjects 1-12) Concentrations of 11-COOH THC in Urine Following Administration of CBME

■ Sublingual   ▨ Buccal   ☐ Oro-Pharngeal   ☐ Oral

NB. Mean data taken from Table 3

TABLE 7. Summary of Plasma THC Pharmacokinetic Parameters–PASS Sublingual

Statistic	$AUC_{0-\infty}$ (ng/ml·min)	AUC_{0-t} (ng/ml·min)	C_{max} (ng/ml)	T_{max} (min)	$t_{1/2}$ (min)
N	12	12	12	12	12
Arithmetic mean	837.3	808.8	5.54	97.5	105.7
Geometric mean	772.5	745	4.64	-	-
Minimum	429.3	406.9	1.14	60	45.4
Maximum	1857.5	1812	12.13	180	193.7
SD	387.34	378.36	3.346	35.32	39.743
CV%	46.3	46.8	60.4	36.2	37.6
Log transformed:					
Mean	6.6496	6.6134	1.5349	-	-
SD	0.4049	0.4087	0.651	-	-

TABLE 8. Summary of Plasma CBD Pharmacokinetic Parameters–PASS Sublingual

Statistic	$AUC_{0-\infty}$ (ng/ml·min)	AUC_{0-t} (ng/ml·min)	C_{max} (ng/ml)	T_{max} (min)	$t_{1/2}$ (min)
N	12	12	12	12	12
Arithmetic mean	427.3	408.5	2.5	97.5	86.35
Geometric mean	370.1	344.3	1.87	-	-
Minimum	137.5	93.8	0.27	45	44.2
Maximum	1106.4	1083.8	6.55	180	201.6
SD	258.86	259.86	1.8281	40.7	47.18
CV%	60.6	63.6	73.2	41.7	54.6
Log transformed:					
Mean	5.9137	5.8414	0.6286	-	-
SD	0.5537	0.625	0.866	-	-

TABLE 9. Summary of Plasma THC Pharmacokinetic Parameters–PASS Oro-Pharyngeal

Statistic	$AUC_{0-\infty}$ (ng/ml·min)	AUC_{0-t} (ng/ml·min)	C_{max} (ng/ml)	T_{max} (min)	$t_{1/2}$ (min)
N	12	12	12	12	12
Arithmetic mean	985.1	962.7	6.11	133.8	81.2
Geometric mean	897.7	874.2	5.06	-	-
Minimum	413.3	404.4	1.94	45	41.8
Maximum	1772.1	1758.3	15.68	300	162.5
SD	440.27	439.99	3.998	91.23	30.838
CV%	44.7	45.7	65.5	68.2	38
Log transformed:					
Mean	6.7998	6.7733	1.621	-	-
SD	0.4545	0.4618	0.648	-	-

TABLE 10. Summary of Plasma CBD Pharmacokinetic Parameters–PASS Oro-Pharyngeal

Statistic	$AUC_{0-\infty}$ (ng/ml·min)	AUC_{0-t} (ng/ml·min)	C_{max} (ng/ml)	T_{max} (min)	$t_{1/2}$ (min)
N	12	12	12	12	12
Arithmetic mean	497	469.1	2.61	122.5	105.5
Geometric mean	417.3	387.8	2.01	-	-
Minimum	128.2	109.4	0.41	45	41.4
Maximum	1286.8	1201.3	6.36	300	186.1
SD	319.34	307.78	1.907	67.94	47.879
CV%	64.3	65.6	73	55.5	45.4
Log transformed:					
Mean	6.0337	5.9606	0.7004	-	-
SD	0.6238	0.657	0.7923	-	-

TABLE 11. Summary of Plasma THC Pharmacokinetic Parameters–PASS Buccal

Statistic	$AUC_{0-\infty}$ (ng/ml·min)	AUC_{0-t} (ng/ml·min)	C_{max} (ng/ml)	T_{max} (min)	$t_{1/2}$ (min)
N	12	12	12	12	12
Arithmetic mean	770.6	751.2	6.14	143.8	80.47
Geometric mean	664.6	640.4	4.39	-	-
Minimum	233.6	225.3	0.88	60	44.6
Maximum	1666.9	1656	19.78	270	168.4
SD	427.22	431.19	5.367	65.06	38.807
CV%	55.4	57.4	87.4	45.3	48.2
Log transformed:					
Mean	6.4992	6.4621	1.4791	-	-
SD	0.5852	0.6081	0.8827	-	-

TABLE 12. Summary of Plasma CBD Pharmacokinetic Parameters–PASS Buccal

Statistic	$AUC_{0-\infty}$ (ng/ml·min)	AUC_{0-t} (ng/ml·min)	C_{max} (ng/ml)	T_{max} (min)	$t_{1/2}$ (min)
N	12	12	12	12	12
Arithmetic mean	407.8	384.1	3.02	167.5	108.39
Geometric mean	328.1	287.9	1.82	-	-
Minimum	100.5	80.4	0.29	60	38.2
Maximum	862.7	852.4	9.91	270	451.4
SD	267.8	277.34	3.1478	78.81	122.936
CV%	65.7	72.2	104.1	47.1	113.4
Log transformed:					
Mean	5.7932	5.6625	0.5996	-	-
SD	0.7146	0.8429	1.0925	-	-

TABLE 13. Summary of Plasma 11-Hydroxy-THC Pharmacokinetic Parameters–PASS Sublingual

Statistic	$AUC_{0-\infty}$ (ng/ml·min)	AUC_{0-t} (ng/ml·min)	C_{max} (ng/ml)	T_{max} (min)	$t_{1/2}$ (min)
N	12	12	12	12	12
Arithmetic mean	1632.5	1522.1	6.24	95	128.84
Geometric mean	1508.2	1410.6	5.7	-	-
Minimum	635.7	621.6	2.67	60	54.3
Maximum	3058.3	2906.3	10.77	165	270.3
SD	687.19	638.68	2.744	26.63	59.252
CV%	42.1	42	43.9	28	46
Log transformed:					
Mean	7.3187	7.2518	1.7409	-	-
SD	0.4198	0.4079	0.45	-	-

TABLE 14. Summary of Plasma 11-Hydroxy-THC Pharmacokinetic Parameters–PASS Buccal

Statistic	$AUC_{0-\infty}$ (ng/ml · min)	AUC_{0-t} (ng/ml · min)	C_{max} (ng/ml)	T_{max} (min)	$t_{1/2}$ (min)
N	12	12	12	12	12
Arithmetic mean	1362.1	1293.2	6.13	143.8	114.34
Geometric mean	1191.5	1123.2	5.48	-	-
Minimum	357.1	345.1	1.83	60	66.4
Maximum	3308.9	3152.3	11.25	270	323.5
SD	753.7	728.83	2.878	69.91	74.866
CV%	55.3	56.4	46.9	48.6	65.5
Log transformed:					
Mean	7.083	7.0239	1.7002	-	-
SD	0.5582	0.574	0.524	-	-

TABLE 15. Summary of Plasma 11-Hydroxy-THC Pharmacokinetic Parameters–PASS Oro-Pharyngeal

Statistic	$AUC_{0-\infty}$ (ng/ml · min)	AUC_{0-t} (ng/ml · min)	C_{max} (ng/ml)	T_{max} (min)	$t_{1/2}$ (min)
N	12	12	12	12	12
Arithmetic mean	1580.3	1477.8	6.45	143.8	125.78
Geometric mean	1520.1	1420.6	5.94	-	-
Minimum	737	688.7	2.95	75	59
Maximum	2483.7	2379.3	13.49	300	260.8
SD	440.08	420.39	2.905	73.05	56.496
CV%	27.8	28.4	45.1	50.8	44.9
Log transformed:					
Mean	7.3265	7.2588	1.7815	-	-
SD	0.3019	0.3032	0.416	-	-

buccal than either sublingual or oro-pharyngeal dosing. The ratios of AUC_{0-t} of 11-hydroxy-THC to THC were 1.5, 1.7 and 1.9:1 (calculated from Table 4) following oro-pharyngeal, buccal and sublingual dosing, respectively.

Inter-subject variability in pharmacokinetics was considerable with CV% of the order of 45 to 70% in AUC, 38 to 68% in T_{max} and 44 to 113% in C_{max} (calculated from Table 4). Following each treatment differences between the lowest and highest C_{max} values observed in individual subjects ranged from 8 to 46-fold, with the range being generally greater for CBD than THC. The difference between lowest and highest AUC_{0-t} was less, being of the order of 11-fold for CBD after all formulations and 4 to 7-fold for THC. While some individuals tended to show consistency in high or low AUC or C_{max} values across all treatments, others showed considerable intra-subject variability.

Analysis of Oral Capsule Pharmacokinetic Parameters

Following administration of the oral capsules the mean T_{max} of CBD was 76 min (Table 16) and for THC 63 min (Table 17). The C_{max} of CBD was 2.47 ng/ml and C_{max} of THC was 6.35 ng/ml. T_{max} of CBD, THC and 11-hydroxy-THC (Table 18) occurred earlier following dosing with oral capsules than dosing with sublingual buccal or oro-pharyngeal sprays.

TABLE 16. Summary of Plasma CBD Pharmacokinetic Parameters–Oral Capsule

Statistic	$AUC_{0-\infty}$ (ng/ml·min)	AUC_{0-t} (ng/ml·min)	C_{max} (ng/ml)	T_{max} (min)	$t_{1/2}$ (min)
N	12	12	12	12	12
Arithmetic mean	362	345.7	2.47	76.3	65.41
Geometric mean	259.1	240.8	1.72	-	-
Minimum	79.1	67.3	0.47	30	22.9
Maximum	932.8	921.1	7.55	180	108.5
SD	298.28	296.28	2.233	50.55	27.58
CV%	82.4	85.7	90.3	66.3	42.2
Log transformed:					
Mean	5.5571	5.4838	0.5406	-	-
SD	0.8779	0.9129	0.8964	-	-

TABLE 17. Summary of Plasma THC Pharmacokinetic Parameters–Oral Capsule

Statistic	$AUC_{0-\infty}$ (ng/ml·min)	AUC_{0-t} (ng/ml·min)	C_{max} (ng/ml)	T_{max} (min)	$t_{1/2}$ (min)
N	12	12	12	12	12
Arithmetic mean	724.8	705.4	6.35	62.5	71.72
Geometric mean	656.2	635	5.79	-	-
Minimum	366	357.8	3.04	30	36.8
Maximum	1744.4	1731.8	14.55	165	134.1
SD	375.66	377.07	3.122	38.82	25.583
CV%	51.8	53.5	49.2	62.1	35.7
Log transformed:					
Mean	6.4864	6.4537	1.7564	-	-
SD	0.45	0.4619	0.4327	-	-

TABLE 18. Summary of Plasma 11-Hydroxy-THC Pharmacokinetic Parameters–Oral Capsule

Statistic	$AUC_{0-\infty}$ (ng/ml·min)	AUC_{0-t} (ng/ml·min)	C_{max} (ng/ml)	T_{max} (min)	$t_{1/2}$ (min)
N	12	12	12	12	12
Arithmetic mean	1480.4	1411	7.87	81.3	100.1
Geometric mean	1394.5	1331.6	7.4	-	-
Minimum	623.6	608.9	4.79	45	67.1
Maximum	2470.5	2389.4	13.64	180	132.4
SD	515.87	487.29	2.958	38.09	17.69
CV%	34.8	34.5	37.6	46.9	17.7
Log transformed:					
Mean	7.2403	7.1941	2.0021	-	-
SD	0.3724	0.3655	0.3585	-	-

Mean AUC_{0-t} and $AUC_{0-\infty}$ of CBD (345.68 and 362.04 ng/ml.min, respectively) were lower, whereas the mean AUC_{0-t} and $AUC_{0-\infty}$ of THC (705.38 and 724.79 ng/ml.min, respectively) were greater following dosing with oral capsules than with the sublingual, buccal or oro-pharyngeal sprays. The bioavailability of THC was approximately twice that of CBD. The mean T_{max} of 11-hydroxy-THC (81 min) was a little later than that of CBD or THC, though still earlier than following dosing with sublingual buccal or oro-pharyngeal sprays. The C_{max} (7.87 ng/ml) for 11-hydroxy-THC was greater than that of THC. Mean AUC_{0-t} and $AUC_{0-\infty}$ (1410.99 and 1480.39 ng/ml.min, respectively) were twice the corresponding values for THC.

Analysis of Safety Parameters

For each of the blood pressure and pulse parameters descriptive statistics (n, mean, SD, median, minimum and maximum) were calculated and summarised at each time point by treatment group. In addition, the calculations were performed for the absolute change from pre-dose.

For each of the ECG parameters (heart rate, PR interval, QT_c interval and QRS width), descriptive statistics (N, mean, SD, median, minimum and maximum) were calculated and summarised at each time point by treatment group. In addition, the calculations were performed for the absolute change from pre-dose. For QT_c, absolute values and changes from pre-dose were categorised as borderline, normal, prolonged according to CPMP guidelines.

Statistical/Analytical Issues

There were no specific statistical or analytical issues in this study.

Plasma Concentration Conclusions

Mean data indicate an almost simultaneous appearance of all three cannabinoids in plasma at 30 minutes after dosing, though in individuals there was considerable variability in the time to first appearance of the cannabinoids (range 15-105 minutes).

Concentrations of THC were higher than the corresponding levels of CBD at most time points. Concentrations of 11-hydroxy-THC exceeded the corresponding concentration of THC at most time points after 45 min. By 720 min (12 h) post-dose, mean concentrations of each cannabinoid were still above the LLOQ.

There was a high degree of inter-subject and intra-subject variability in the plasma concentrations achieved.

Urine Concentration Conclusions

No statistical analyses were carried out on the urine data. Urine samples were collected in polypropylene containers and due to the affinity of cannabinoids to plastic, the accuracy of the urine data is not known. 11-COOH THC (a metabolite of THC) was detected in urine throughout the sampling period in quantifiable amounts.

The excretion of 11-COOH THC began within the first 0.5 to 1 hour after dosing, peaked during the 3-6 h collection period and thereafter decreased. Administration of the oral capsules resulted in resulted in the greatest total concentrations of 11-COOH THC excreted, followed by dosing sublingually and the buccal and oro-pharyngeal routes showed approximately the same extent of excretion of 11-COOH THC throughout the sampling period.

Pharmacokinetic Conclusions

T_{max} of CBD and THC occurred earlier following sublingual administration than oro-pharyngeal or buccal although only the difference in T_{max} of CBD compared with buccal was statistically significant.

C_{max} of both CBD and THC for the PASS test treatments was greatest following buccal administration although this was not statistically significant. AUC was greatest following oro-pharyngeal administration and was statistically significantly greater than following buccal administration. The lower bioavailability, as measured by AUC, following buccal administration when compared to the sublingual and oro-pharyngeal routes may be related to the difficulty of spraying onto the inside of the cheek reported during the study. Buccal administration of the PASS test treatment resulted in a later T_{max} but greater C_{max} when compared to the sublingual and oro-pharyngeal routes.

Comparison of the sublingual and oro-pharyngeal routes showed no statistically significant difference in THC or CBD pharmacokinetic parameters measured.

Pharmacokinetic parameters following administration of the oral capsule were not statistically compared to the other routes as this was an early investigation into the safety and tolerability of this dose route. However, this dosage form and route of administration appeared to show an early T_{max} of both CBD and THC. Mean C_{max} of THC and 11-hydroxy-THC were greater, but in contrast the C_{max} of CBD was lower, than following the PASS treatments.

Relative to THC, the plasma level AUC of 11-hydroxy-THC was proportionally greatest following dosing with the oral capsules which could be a reflection of greater metabolism by this route. Of the PASS treatments the ratio of 11-hydroxy-THC to THC was greatest following sublingual and least following oro-pharyngeal dosing.

The oral capsule has good bioavailability, and provided, as is the case here, the formulation is not oil based, may be a viable formulation when self-titration is not necessary. There was very wide inter- and to a lesser extent intra-subject variability in pharmacokinetics. Differences in mean values between the routes of administration, even when statistically significant, are small relative to the very wide range of values between subjects.

Safety Evaluation

The test treatments were well tolerated by all subjects with no Serious Adverse Events (SAEs) recorded throughout the study and no subject withdrawals. Peak concentrations of cannabinoids in plasma did not correspond with AEs or other events.

Adverse Events

A summary of treatment-emergent/treatment-related AEs is presented in Table 19. A total of 146 AEs occurred in 12 subjects through-

TABLE 19. Summary of Subjects Who Experienced Treatment Emergent, Treatment Related Adverse Events

Event	Treatment			
	A	B	C	D
No. of subjects with ≥ 1 event	12 (100%)	12 (100%)	11 (91.7%)	10 (83.3%)
Cardiac disorders	2	2	1	2
Palpitations	2			
Sinus tachycardia	1	2	1	2
Gastrointestinal disorders	6	6	6	2
Aptyalism				2
Throat irritation	6	6	6	
General disorders and administration site conditions	4	7	6	5
Application site irritation	3	3	4	
Feeling cold		1	1	1
Feeling of relaxation	1	2	1	3
Lethargy	1	1	2	1
Injury, poisoning and procedural complications	2	1	3	2
Drug toxicity NOS	2	1	3	2
Nervous system disorders	9	10	9	9
Coordination abnormal NOS				1
Disturbance in attention	1		1	3
Dizziness	7	5	8	7
Dysgeusia	1	1		
Headache NOS	1	3	1	1
Paraesthesia	2	2		3
Paraesthesia oral NOS	1	3		1
Somnolence	4	3	2	5
Psychiatric disorders	2	2	1	1
Anxiety NEC	1	1		
Dissociation	1	1	1	
Restlessness	1	2	1	1
Skin and subcutaneous tissue disorders	1	0	0	0
Rash maculo-papular	1			

Note: treatment related = definitely, probably, possibly related

out the study. Two events were classified as moderate (flu-like illness and pharyngeal irritation) and the remaining 144 were classified as mild. Three events were classified as not related to test treatment, (flu-like illness, coryza and feels cold), leaving 143 considered to be possibly, probably or definitely related to the test treatment. At the end of the study, all the events, with the exception of maculo-papular rash of the neck and shoulders, had resolved without treatment. The maculo-papular rash did not require treatment and follow up was continued at the clinical site until resolution.

The most common AEs experienced were dizziness, throat irritation, somnolence, and application site irritation.

The number of subjects experiencing treatment related throat irritation were the same for the sublingual (6), buccal (6) and oro-pharyngeal (6) routes, however there were none reported for the oral capsule. Treatment related application site irritation was experienced with the sublingual (3), buccal (3) and oro-pharyngeal (4), however no application site irritation AEs were reported for the oral capsules. Treatment related paraesthesia was experienced after dosing sublingually (2), buccally (2) and with the oral capsule (3), however no paraesthesia AEs were reported in the subjects receiving PASS oro-pharyngeally.

There were no deaths or SAEs during the study, and no withdrawals due to AEs.

Clinical Laboratory Evaluation

There were no clinically significant changes in the individual or mean haematology or clinical chemistry parameters from pre-dose to post-study (Table 5). There were no haematology or clinical chemistry parameter results (or changes from pre-study to post-study) observed, which were considered to be clinically significant. There were no clinically significant individual subject changes in any safety parameters noted throughout the study. There were no results observed or reported throughout the study that were considered by the investigator to be clinically significant abnormal results.

Vital Signs, Physical Findings and Other Observations Related to Safety

There were no changes in vital signs, physical findings or other safety analyses recorded throughout the study that were considered by the investigator to be clinically significant.

Safety Conclusions

All test treatments were well tolerated by all subjects with no SAEs occurring throughout the study. Most of the AEs experienced by subjects were mild and resolved without treatment. The most common AEs experienced across all test treatments were dizziness, throat irritation, somnolence, and application site irritation. The only notable differences in AEs between test treatment groups were throat irritation and application site irritation, which were not seen with the oral capsule, and paraesthesia which was not seen with oro-pharyngeal dosing.

DISCUSSION AND OVERALL CONCLUSIONS

All routes of administration were well tolerated by all subjects with no SAEs and no withdrawals due to AEs.

There was a wide intra-subject variability in each of the pharmacokinetic parameters. This variation may be due to many factors such as amount of dose swallowed instead of absorption through the oral mucosa, breakfast on the morning of dosing, or levels of exercise undertaken by each subject.

By 720 min (12 h) post-dose mean concentrations of each cannabinoid were still above the LLOQ, indicating that redistribution within the body may still be occurring. The sublingual and oro-pharyngeal routes of administration appear to have the same pharmacokinetic results. The buccal pharmacokinetic parameters are lower when compared to the sublingual and oro-pharyngeal routes. Overall, the results indicate that administration of the liquid spray (GW-1000-02) need not be limited to sublingual administration.

The oral capsule has good bioavailability and provided as is the case here, the formulation is not oil based, may be a viable formulation when self-titration is not necessary. The urine samples were collected in polypropylene containers therefore the reliability of the urine concentration data is not known. Excretion in urine for all four test treatments showed a similar pattern with excretion in significant amounts beginning as the concentrations of THC and 11-hydroxy-THC in plasma were decreasing. This suggests that a portion of the cannabinoids are rapidly metabolised and excreted via the kidneys and are not re-distributed to body tissues such as adipose tissue. In some subjects' excretion of 11-COOH-THC was still occurring pre-next dose suggesting that a portion of the test treatment is re-distributed to body tissues and slowly eliminated via the kidneys. During this slow elimination phase (12 hours to six days), no CBD, THC or 11-hydroxy-THC can be detected in plasma suggesting that after a six day washout period either all THC is metabolised to 11-COOH THC or is re-distributed to other body tissues.

REFERENCES

Agurell, S., M. Halldin, J. E. Lindgren, A. Ohlsson, M. Widman, H. Gillespie, and L. Hollister. 1986. Pharmacokinetics and metabolism of delta 1-tetrahydrocannabinol and other cannabinoids with emphasis on man. *Pharmacol Rev* 38 (1):21-43.

British Medical Association. 1997. *Therapeutic uses of cannabis.* Amsterdam: Harwood Academic Publishers.

Baker, P. B., T. A. Gough, and B. J. Taylor. 1983. The physical and chemical features of Cannabis plants grown in the United Kingdom of Great Britain and Northern Ireland from seeds of known origin–Part II: Second generation studies. *Bull Narc* 35 (1):51-62.

GW Pharmaceuticals. 2002. Investigators brochure for cannabis based medicine extract (CBME). Porton Down: GW Pharmaceuticals.

House of Lords. 1998. Cannabis: The scientific and medical evidence. London: House of Lords Select Committee on Science and Technology, Stationery Office.

Maykut, M. O. 1985. Health consequences of acute and chronic marihuana use. *Prog Neuropsychopharmacol Biol Psychiatry* 9 (3):209-38.

McPartland, J. M., and E. B. Russo. 2001. Cannabis and cannabis extracts: Greater than the sum of their parts? *J Cannabis Therapeutics* 1 (3-4):103-132.

Portenoy, R. K. 1990. Chronic opioid therapy in nonmalignant pain. *J Pain Symptom Manage* 5 (1 Suppl):S46-62.

Porter, J., and H. Jick. 1980. Addiction rare in patients treated with narcotics. *N Engl J Med* 302 (2):123.

Whittle, B. A., G. W. Guy, and P. Robson. 2001. Prospects for new cannabis-based prescription medicines. *J Cannabis Therapeutics* 1 (3-4):183-205.

A Phase I, Double Blind, Three-Way Crossover Study to Assess the Pharmacokinetic Profile of Cannabis Based Medicine Extract (CBME) Administered Sublingually in Variant Cannabinoid Ratios in Normal Healthy Male Volunteers (GWPK0215)

G. W. Guy
P. J. Robson

SUMMARY. Primary objectives of this study were to assess the pharmacokinetic characteristics of CBME when administered sublingually in different ratios, to determine if the pharmacokinetic profiles of THC and its metabolite 11-hydroxy-THC are different when administered sublingually in different formulations, and to characterise the pharmacokinetic profile of CBD when administered with THC in equal amounts.

G. W. Guy and P. J. Robson are affiliated with GW Pharmaceuticals plc, Porton Down Science Park, Salisbury, Wiltshire, SP4 0JQ, UK.

[Haworth co-indexing entry note]: "A Phase I, Double Blind, Three-Way Crossover Study to Assess the Pharmacokinetic Profile of Cannabis Based Medicine Extract (CBME) Administered Sublingually in Variant Cannabinoid Ratios in Normal Healthy Male Volunteers (GWPK0215)." Guy, G. W., and P. J. Robson. Co-published simultaneously in *Journal of Cannabis Therapeutics* (The Haworth Integrative Healing Press, an imprint of The Haworth Press, Inc.) Vol. 3, No. 4, 2003, pp. 121-152; and: *Cannabis: From Pariah to Prescription* (ed: Ethan Russo) The Haworth Integrative Healing Press, an imprint of The Haworth Press, Inc., 2003, pp. 121-152. Single or multiple copies of this article are available for a fee from The Haworth Document Delivery Service [1-800-HAWORTH, 9:00 a.m. - 5:00 p.m. (EST). E-mail address: docdelivery@haworthpress.com].

http://www.haworthpress.com/store/product.asp?sku=J175
© 2003 by The Haworth Press, Inc. All rights reserved.
10.1300/J175v03n04_02

Secondary objectives were to determine if there was a correlation between intoxication levels and plasma concentrations of THC and/or its metabolite 11-hydroxy-THC, and to assess safety and tolerability of CBME when administered sublingually.

Methodology employed a double-blind, randomised, three-way crossover study of placebo, High THC and CBD:THC administered sublingually as a liquid spray. Twenty-four subjects were planned, dosed, completed the study and were analysed.

Test products were Δ^9-tetrahydrocannabinol (THC, formulated as 25 mg THC per ml) with or without cannabidiol (CBD) (formulated as 25 mg CBD + 25 mg THC per ml) formulated in ethanol (Eth):propylene glycol (PG) with peppermint (ppmt) flavouring or matching placebo, administered with a 100 µl pump. Each subject received one single dose of 10 mg THC and one single dose of 10 mg CBD + 10 mg THC plus a single dose of placebo in a randomised manner on three separate occasions. The washout period was six days between each dose. Placebo was Eth:PG in a 50:50 ratio with ppmt flavouring, administered with a 100 µl actuator pump.

Mean plasma concentrations show that following administration of both High THC and CBD:THC formulations CBD and or THC was detectable in plasma in measurable concentrations 15-30 minutes after dosing, although individual subjects showed quite wide variability, 15 to 135 minutes, to appearance measurable concentrations. At all time points up to 180 minutes after dosing mean concentrations of THC were greater following the High THC formulation than CBD:THC. Concentrations of THC were also greater than corresponding concentrations of CBD following the CBD:THC treatment.

There were no statistically significant differences in mean C_{max}, $t_{1/2}$, AUC_{0-t} and $AUC_{0-\infty}$ of both THC and 11-hydroxy-THC between the High THC and CBD:THC formulations. THC T_{max} was statistically significantly later following CBD:THC than High THC ($p = 0.014$) and this was the only statistically significant difference in pharmacokinetic parameters between the treatments. The AUC values (AUC_{0-t} and $AUC_{0-\infty}$) for THC show an approximate 8 to 10-fold difference between the lowest and highest subject values while the difference for CBD was approximately 3.5 to 4-fold. Differences in C_{max} were 20 to 30 fold for THC and approximately 14-fold for CBD. Intra-subject differences in values for THC between treatments were smaller though differences in C_{max} of up to 5-fold and 3-fold in AUC (AUC_{0-t} and $AUC_{0-\infty}$) were observed. Other than a single isolated significant difference in T_{max} there were no significant differences in pharmacokinetic parameters between the CBD:THC and High THC formulations. The bioavailability of THC appears to be greater than that of CBD.

Mean intoxication scores on both CBME treatments were very low throughout the observation period. The majority of subjects scored zero for the majority of assessment points and there were few scores greater than three on the Box Scale 11 (BS-11). Recorded intoxication scores do not seem to show a direct relationship to plasma concentrations of THC and/or 11-hydroxy-THC either within or between subjects. The time of intoxication scores in individual subjects do not seem to relate consistently with the timing of increases in plasma concentrations or maximal concentrations of THC or 11-hydroxy-THC. Neither is there an apparent relationship between subjects reporting intoxication and those with the highest plasma levels of THC or 11-hydroxy-THC.

No subjects withdrew from the study as a result of adverse events and both active and the placebo test treatments were well tolerated. The treatment with the least number of treatment related adverse events was placebo. High THC and CBD:THC had a greater number of subjects who experienced intoxication type adverse events and application site type reactions. The most common overall adverse event experienced was throat irritation, followed by dizziness, somnolence, oral paraesthesia and then headache. All the events were mild and only two events needed any treatment. There were no clinically significant changes from baseline for haematology, biochemistry, vital signs or ECGs.

There was wide inter- and intra-subject variability in pharmacokinetic parameters with up to 10-fold differences in THC AUC between subjects and even greater differences in C_{max}. Results suggest that there are no overall statistically significant differences between the pharmacokinetic parameters of High THC and CBD:THC other than a delay in T_{max}. Considering the wide inter- and intra-subject variability in pharmacokinetic parameters including T_{max} this is unlikely to be clinically important in a medication that is self titrated by the patient. *[Article copies available for a fee from The Haworth Document Delivery Service: 1-800-HAWORTH. E-mail address: <docdelivery@haworthpress.com> Website: <http://www.HaworthPress. com> © 2003 by The Haworth Press, Inc. All rights reserved.]*

KEYWORDS. Cannabinoids, cannabis, THC, cannabidiol, medical marijuana, pharmacokinetics, pharmacodynamics, multiple sclerosis, botanical extracts, alternative delivery systems, harm reduction

INTRODUCTION

Cannabis plants (*Cannabis sativa*) contain approximately 60 different cannabinoids (Association 1997) and in the UK, oral tinctures of cannabis were prescribed until cannabis was made a Schedule 1 controlled substance in the Misuse of Drugs Act, 1971. The prevalence of recreational cannabis use increased markedly in the UK after 1960, reaching a peak in the late 1970s. This resulted in a large number of individuals with a range of intractable medical disorders being exposed to the drug, and many of these discovered that cannabis could apparently relieve symptoms not alleviated by standard treatments. This was strikingly the case with certain neurological disorders, particularly multiple sclerosis (MS). The black market cannabis available to those patients is thought to have contained approximately equal amounts of the cannabinoids Δ^9-tetrahydrocannabinol (THC) and cannabidiol (CBD) (Baker, Gough, and Taylor 1983). The importance of CBD lies not only in its own inherent therapeutic profile but also in its ability to modulate some of the undesirable effects of THC through both pharmacokinetic and pharmacodynamic mechanisms (McPartland and Russo 2001). MS patients claimed beneficial effects from cannabis in many core symptoms, including pain, urinary disturbance, tremor, spasm and spasticity (Association 1997). The MS Society estimated in 1998 that up to 4% (3,400) of UK MS sufferers used cannabis medicinally (Lords 1998).

Cannabinoid clinical research has often focussed on synthetic analogues of THC, the principal psychoactive cannabinoid, given orally. This has not taken the possible therapeutic contribution of the other cannabinoid and non-cannabinoid plant components into account, or the slow and unpredictable absorption of cannabinoids via the gastrointestinal tract (Agurell et al. 1986). Under these conditions it has been difficult to titrate cannabinoids accurately to a therapeutic effect. Research involving plant-derived material has often reported only the THC content (Maykut 1985) of the preparations, making valid comparisons between studies difficult. GW Pharma Ltd (GW) has developed cannabis based medicine extracts (CBMEs) derived from plant cultivars that produce high and reproducible yields of specified cannabinoids. CBMEs contain a defined amount of the specified cannabinoid(s), plus the minor cannabinoids and also terpenes and flavonoids. The specified cannabinoids constitute at least 90% of the total cannabinoid content of the extracts. The minor cannabinoids and other constituents add to the overall therapeutic profile of the CBMEs and may play a role in stabilising the extract (Whittle, Guy, and Robson 2001). Early clinical studies indicated that sublingual dosing with CBME was feasible, well tolerated and convenient for titration. The concept of self-titration was readily understood by patients and worked well in practice. Dosing patterns tended to resemble those seen in the patient controlled analgesia technique used in post-operative pain control; with small doses administered as and when patients require them, up to a maximal rate and daily limit (Pharmaceuticals 2002). The Phase 2 experience has supported some of the wide-range of effects reported anecdotally for cannabis. It has also shown that for most patients the therapeutic benefits of CBMEs could be obtained at doses below those that cause marked intoxication (the 'high'). This is consistent with experience in patients receiving opioids for pain relief, where therapeutic use rarely leads to misuse (Porter and Jick 1980; Portenoy 1990). Onset of intoxication may be an indicator of over-titration. However the range of daily dose required is subject to a high inter-individual variability.

SATIVX (1:1 THC:CMD CBME) was administered as an oromucosal spray, and contains an equal proportion of THC and CBD, similar to the cannabinoid profile of the cannabis thought to be most commonly available on the European black market (Porter and Jick 1980; Portenoy 1990). The High-THC CBME was administered as an oromucosal spray, and contains over 90% of cannabinoids as THC. Placebo was administered as sublingual liquid spray and was used as a reference treatment to reduce bias.

GWPK0215 was a Phase I clinical study that primarily aimed to assess the PK profiles of each test treatment. It was also designed to assess safety and tolerability of the test treatments.

Primary objectives of this study were to assess the PK characteristics of CBME when administered sublingually in different ratios, to determine if the PK profiles of THC and its metabolite, 11-hydroxy-THC, are different when administered sublingually in different formulations, and to characterise the PK profile of CBD when administered with THC in equal amounts. Secondary objectives were to determine if there is a correlation between intoxication levels and plasma concentrations of THC and/or its metabolite 11-hydroxy-THC, and also to assess safety and tolerability of CBME when administered sublingually.

OVERALL STUDY DESIGN AND PLAN–DESCRIPTION

The study was a double-blind, three-period, three-way randomised crossover using single doses of 10 mg THC, 10 mg CBD + 10 mg THC and placebo. The test treatment was administered sublingually as a liquid spray according to the pre-determined randomisation scheme. The washout period between each dose was six days.

High THC CBME was formulated in 50% ethanol (Eth), 50% propylene glycol (PG) at a concentration of 25 mg THC per ml of Eth:PG with peppermint flavouring. It was delivered via pump action spray at 100 µl per actuation

SATIVEX (1:1 THC:CBD CBME) was formulated in 50% ethanol (Eth), 50% propylene glycol (PG) at a concentration of 25 mg CBD + 25 mg THC per ml of Eth:PG with peppermint flavouring. It was delivered via pump action spray at 100 µl per actuation. Placebo was formulated as Eth:PG in a 50:50 ratio with peppermint flavouring delivered via pump action spray at 100 µl per actuation.

Subjects were required to undergo a pre-study screen no more than 21 days prior to first dose administration to determine their eligibility to take part in the study. Only those subjects who were healthy and complied with all the study requirements were deemed eligible for participation.

These test treatments were chosen as they were the formulation and treatments that were used in the GW Pharmaceuticals clinical programme. The dose administered in this study (10 mg CBD and/or 10 mg THC) was chosen as this is a high single dose of the test treatment when used by patients in a self-titrated regime and is known to be well tolerated by normal healthy subjects.

A randomised cross-over design was chosen to enable both inter- and intra-subject comparisons of PK and pharmacodynamic data and to reduce period effect. The study was double-blind to ensure no bias could be introduced when assessing adverse events (AEs) and pharmacodynamic effects.

A six-day washout was chosen to ensure all cannabinoids were below the limit of quantification and eased the scheduling of the study in the clinical unit.

GW specified that only subjects with previous experience with the effects of cannabis be included in this trial to ensure that subjects recognised the adverse effects (in particular the 'recreational high') they may experience as a result of being dosed with the test treatments.

For inclusion in the study subjects were required to fulfil ALL of the following criteria:

i. Adult male aged between 18 and 50 years and BMI of between 19 and 30 kg/m^2.
ii. Had given written informed consent.
iii. Had experienced the effects of cannabis more than once.

Subjects were deemed not acceptable for participation in the study if any of the following criteria applied:

    i. Had a presence of cardiovascular, haematological, hepatic, gastro-intestinal, renal, pulmonary, neurological or psychiatric disease.
    ii. Had a history or presence of schizophrenic-type illness.
    iii. Had a history or presence of drug or alcohol abuse in the past 12 months.
    iv. Had been hospitalised in the three months prior to dosing.
    v. Had lost or donated > 400 ml of blood in the three months prior to dosing.
    vi. Had participated in a clinical trial in the three months prior to dosing.
    vii. Had a history or presence of allergies to cannabis and/or its metabolites.
    viii. Were taking or had taken a course of prescribed medication in the four weeks prior to dosing.
    ix. Were taking or had taken over-the-counter medication, excluding paracetamol and/or vitamins but including mega dose vitamin therapy, within the week before administration of the first dose.
    x. Had blood and/or urinalysis results at screening, which, in the opinion of the Principal Investigator were clinically significant.
    xi. Had a resting blood pressure (BP) of > 150/90 mmHg or < 90/50 mmHg and a pulse of > 100 beats per minute (BPM) or < 40 BPM.
    xii. Had an ECG which, in the opinion of the Principal Investigator was clinically significant.
    xiii. Smoked ≥ 5 cigarettes or used the equivalent in tobacco per day.
    xiv. Regularly consumed > 28 units of alcohol per week.

Subjects were required to agree to the following:

    i. Using barrier methods of contraception during and for three months after completion of the study.
    ii. Abstaining from consuming all foods and beverages containing caffeine and/or alcohol for 36 h before until the end of each confinement period.
    iii. Abstaining from taking any medications (prescription and/or over-the-counter) and drugs, for the duration of the study.

iv. Not smoking or using tobacco products during each confinement period.
v. Not donating blood in the three months after completion of the study.
vi. Not participating in another clinical trial for 3 months after completion of this one.

The subjects were free to withdraw from the study without explanation at any time and without prejudice to future medical care. Subjects may have been withdrawn from the study at any time if it was considered to be in the best interest of the subject's safety.

A single dose of 10 mg THC, a single dose of 10 mg CBD + 10 mg THC and a single dose of placebo were administered sublingually to each of 24 subjects on three separate occasions in a randomised manner. Each single dose consisted of a series of four actuations of 100 µl volume each (2.5 mg CBD and/or 2.5 mg THC per actuation) and each actuation was administered five minutes apart. Each subject received all of the test treatments once. Each vial was identified with no less than study number, subject number, period number, batch number and expiry date.

Subjects were randomised to a dose sequence using a Williams Square Design provided by GW. All subjects were randomised to receive a single dose of each of the test treatments once in each of the three periods.

The dosing regime and doses chosen are well tolerated by both subjects and patients. The dose given has been previously used in other GW Phase I studies and has been shown to produce both quantifiable drug concentrations in plasma and pharmacodynamic effects.

The subjects were dosed in three groups of eight subjects (Group 1; Subjects 101-108, Group 2; Subjects 109-116 and Group 3; Subjects 117-124). The test treatments were administered in the morning of each dosing day according to the randomisation scheme. Subjects were dosed in the morning to allow blood samples to be taken and procedures to be carried out up to 24 h post-dose with minimal disruption to the subjects during the night. A minimum of six days washout between each dose was specified as previous data and drug of abuse screens have indicated that concentrations of each cannabinoid from a single dose of CBME are below the limit of quantification by this time.

The study was double-blind. Unblinding envelopes were retained at the study site and a duplicate set was retained at GW. All subjects completed the study without experiencing any serious adverse events (SAEs) and unblinding was not required. Upon completion of the in-life phase of the study all unblinding envelopes were returned to GW intact.

Only one subject (Subject 101) took medication during the study.

Subjects were dosed by the Principal Investigator or suitably trained designee. Subjects were instructed to allow each spray of the study formulation to absorb under their tongue and not to swallow, if possible, until the drug had been absorbed. The actual time of administration of each actuation was recorded in the CRF (Case Report Form) and the dosing procedure was witnessed by a dose verifier. All subjects received all of the scheduled doses and there were no deviations from the dosing regimen.

Only those subjects who were healthy and complied with all the study requirements were deemed eligible for participation. The screening procedures comprised the following assessments/measurements: The subjects' date of birth, sex, race, height, weight, BMI, previous cannabis experience, tobacco and alcohol habits were recorded. Subjects were asked to provide details of any drugs, vitamins or medications they had taken in the four weeks prior to screening or were taking at the time of screening.

Details of their previous medical history were also recorded. Subjects underwent a physical examination to determine if there were any abnormalities in any body systems. BP (systolic/diastolic) and pulse were measured after the subject had been seated for no less than 2 minutes. Oral temperature was also measured. A 12-lead ECG (electrocardiogram) was taken for each subject. At least the following ECG parameters were recorded: HR (heart rate), PR, QT_c and QRS intervals. The ECGs were expertly read by Cardio Analytics for ventricular rate, PR interval, QRS duration and QT interval.

Subjects were required to provide a urine sample for routine urinalysis to include protein glucose, ketones, bilirubin, nitrites, blood, urobilinogen, haemoglobin (Hb), and Ph. Microscopy was required to be carried out on any abnormal samples. The samples provided were also screened for alcohol and drugs of abuse, including methadone, benzodiazepines, cocaine, amphetamines, THC, opiates and barbiturates

A blood sample was taken in an EDTA blood tube for full haematology analysis. A blood sample was taken in a gel blood tube for clinical chemistry analysis. The following clinical chemistry parameters were measured: sodium, potassium, urea, creatinine, total bilirubin, alkaline phosphatase, total protein, calcium, gamma glutamyl transferase (GGT), albumin, aspartate aminotransferase (AST), alanine aminotransferase

(ALT). A blood sample was taken in a gel blood tube to screen for the serological presence of past or present Hepatitis B and/or C.

Subjects were required to arrive at the clinic approximately 12 hours prior to dosing for each study period. Each subject's health status was updated and pre-dose procedures (health status update, BP and pulse, alcohol and drug of abuse screen, ECG, Box Scale-11 and blood sample for plasma concentration analysis) were carried out. Only subjects who complied with the requirements of the study were accepted for inclusion in the study.

Blood samples (5 ml) for pharmacokinetic analysis were collected into lithium heparin blood tubes via indwelling cannula or individual venipuncture. Samples were placed immediately into an ice bath until centrifuged (3000 RPM for 10 min at 4°C). The resultant plasma was decanted into two identical pre-labelled silanised amber glass plasma tubes and stored in a freezer at $-20°C$ until shipped to the analytical laboratory.

Blood samples were collected pre-dose and at the following times post start of dosing: 15, 30 and 45 m and 1 h, 1 h 10 m, 1 h 20 m, 1 h 30 m, 1 h 40 m, 1 h 50 m, 2 h, 2 h 15 m, 2 h 30 m, 3, 6, 9, 12 and 24 h post first actuation in each period. Plasma concentrations of CBD, THC and 11-hydroxy-THC were measured in each plasma sample.

SAFETY ASSESSMENTS

Each subject was required to provide a urine sample for a urine drug screen at check in for each dosing period. The drug screen was required to be negative for all drugs pre-dose Period 1. For Periods 2 and 3, positive THC results may have occurred due to administration of test treatment in the previous period and therefore screening for THC was not carried out post Period 1. The urine sample was required to be negative for all other drugs for the subject to be eligible to continue.

The urine sample provided at check-in for each study period for the drug screen was also screened for alcohol. All subjects were required to have a negative alcohol screen to be considered eligible to continue in the study.

12-Lead ECGs were taken for each subject at the following times: pre-dose, 1, 2, 12 and 24 h post-dose. The QT_c intervals for all ECGs were read manually by Cardio Analytics, ITTC Building 2, Tamar Science Park, 1 Davy Road, Derriford, Plymouth, PL6 8BX. Subjects' blood pressure and pulse were measured pre-dose and at 15, 30 and 45 min, 1, 1.5, 2, 3, 6, 9, 12 and 24 h post start of dosing.

Adverse Effects

Subject health was monitored continuously throughout the study for AEs and pharmacodynamic effects and subjects were encouraged to inform the clinical staff of any changes in their health as soon as possible. In addition, subjects' health was monitored by asking non-leading questions pre-dose and at the following times post-dose: 15, 30 and 45 min, 1, 1.5, 2, 2.5, 3, 6, 9, 12 and 24 h post-dose. Any concomitant medications taken during the study were recorded in the subjects CRF.

Box Scale-11 for Intoxication

Subjects were required to complete a Box Scale 11 (BS-11) to describe how intoxicated they were feeling at the following times: pre-dose, 15 m, 30 m, 45 m, 1 h, 1 h 30 m, 2 h, 3 h, 6 h, 9 h, 12 h and 24 h post start of dosing.

Palatability/Dose Questionnaire

As soon as possible after dosing, subjects were asked to complete a questionnaire about the palatability and sensation of the test treatment experienced during and immediately after dosing.

Food and Beverages

A standard low fat breakfast approximately 30 min before dosing for each subject. From 15 min prior to 15 min post-dosing, subjects were required to abstain from consuming food and beverages. Thereafter, decaffeinated beverages and snacks, e.g., digestive biscuits, were available *ad libitum* throughout each confinement period. Subjects were provided with standard meals at approximately 4 and 10 h post-dose (lunch and dinner, respectively) (Table 1).

Check-Out Procedures

After completion of the 24 h study procedures at the end of Periods 1 and 2 and if deemed by the Investigator to be well enough to leave, subjects were discharged from the clinical unit. Prior to discharge, any on-

TABLE 1. Menu

Day 1*			Day 2
Breakfast	Lunch	Dinner	Breakfast
Orange juice	Jacket potato	Chicken	Orange juice
	Cheese	Quorn Fillet (v)	
Cereals	Coleslaw and Salad	Roast Potatoes	Cereals
toast with butter & preserves		peas and carrots	toast with butter & preserves
	Yoghurt	gravy	
		gravy (v)	
tea/coffee**	tea/coffee**		tea/coffee**
orange/lemon	orange/lemon	Peach melba ice cream sundae	orange/lemon
		tea/coffee**	
		orange/lemon	

NB: At each admission (Day -1) subjects were permitted 2 digestive biscuits, decaffeinated tea/coffee and orange/lemon
*Digestive Biscuits and Drinks available throughout the day ** decaffeinated v = vegetarian

going AEs were updated and follow up arranged if required. Prior to Period 3 discharge, subjects were required to undergo a physical examination, blood samples were taken for haematology and clinical chemistry, urinalysis was carried out, a 12-lead ECG was taken and vital signs recorded as per screening. Ongoing AEs were updated and if required arrangements were made to follow up with the subjects after they left the clinical unit.

DATA QUALITY ASSURANCE

Study Monitoring

All details regarding the study were documented within individual CRFs provided by GW for each subject. All data recorded during the study were checked against source data and for compliance with Good Clinical Practice (GCP), internal Standard Operating Procedures (SOPs), working practices and protocol requirements. Monitoring of the study progress and conduct was ongoing throughout the study. Monitoring was conducted by the Clinical Department of GW and was conducted according to GW SOPs. An initiation visit was carried out prior to the start of the study to train site clinical staff on CRF completion, dosing and AE procedures. Training was provided by GW throughout the study as required. Haematology and clinical chemistry analysis were carried out by Leicester General Hospital

Investigator Responsibilities

The Investigator was responsible for monitoring the study conduct to ensure that the rights of the subject were protected, the reported study data were accurate, complete and verifiable and that the conduct of the study was in compliance with ICH GCP. At the end of the study the Principal Investigator reviewed and signed each CRF declaring the data to be true and accurate. If corrections were made after review the Investigator acknowledged the changes by re-signing and dating the CRF.

Clinical Data Management

All study data were collected by GW, who were responsible for evaluation, collation and analysis. Data were subject to quality control procedures. All data were double entered into a Microsoft® Excel 2000 spreadsheet with 10% quality control checks according to GW SOPs. Clinical Quality Audits were carried out by the GW Quality Assurance Department, two Quality Assurance evaluation were carried out and the Pharmacovigilance function was the subject of an internal process audit.

Pharmacokinetic Analysis

All subjects who were dosed and had no more than two missed blood samples were deemed evaluable for, and were included in, PK analyses. All analyses and summary statistics were carried out and derived using SAS v8. All p-values quoted are two-sided. Summary statistics were calculated for each mean PK parameter and treatment (arithmetic mean, N, SD, CV%, minimum, maximum for all parameters and additionally the geometric mean for AUC_{0-t}, $AUC_{0-\infty}$ and C_{max}). AUC_{0-t}, $AUC_{0-\infty}$ and C_{max} were natural log transformed prior to analysis, T_{max} and $t_{1/2}$ were analysed untransformed. For the analytes THC and 11-hydroxy-THC, each parameter was analysed using analysis of variance (ANOVA) with subject and treatment as factors (for High THC and CBD:THC). Least square means are presented for each test treatment. Point estimates of the differences between least square means are presented with the corresponding 95% confidence intervals. For log transformed variables, the contrasts were also back transformed to provide ratios and corresponding 95% confidence intervals. For the analyte CBD which was measurable for only the CBD:THC test treatment, the data are presented descriptively only. K_{el} is presented descriptively only.

Pharmacodynamic Analysis

All subjects who completed at least one study period were evaluable for pharmacodynamic analysis. Intoxication, measured by Box Scale-11, was summarised by treatment group. Means and standard deviations were also calculated.

SAFETY ANALYSIS

Adverse Events

AEs were coded by Medical Dictionary for Regulatory Activities (MedDRA) preferred term and system organ class. These are summarised by test treatment for treatment emergent all causality and treatment emergent treatment related AEs showing the number of subjects with at least one AE and the number of subjects with at least one AE by preferred term within system organ class.

Clinical Laboratory Tests

Laboratory data collected pre- and post-study are summarised descriptively at each of the two time-points and also as the change post-study compared to pre-study.

Concomitant Medications

Subject 101 took concomitant medications between Periods 2 and 3.

Blood Pressure, Pulse and Oral Temperature

Vital signs (pulse, systolic BP, diastolic BP and oral temperature) were monitored. BP and pulse are presented descriptively at each time point up to 12 h post-dose for each test treatment.

12-Lead ECG

ECG parameters (HR, PR interval, QT interval, QT_c and QRS width) were monitored descriptively (N, mean, SD, median, minimum, maximum) pre-study and at each time point up to 12 h post-dose for each test treatment. In addition, QT_c values were classified as either normal, borderline or prolonged. For QT_c, absolute values and changes from pre-dose were categorised as borderline, normal, prolonged according to Committee for Proprietary Medicinal Products (CPMP) guidelines.

Palatability Questionnaire

Each question of the palatability questionnaire was been presented descriptively using frequency tables for each test treatment.

Determination of Sample Size

No formal sample size calculation was carried out for this study. The number of subjects is considered to be sufficient to provide information on the pharmacokinetics of the two formulations.

Changes in the Conduct of the Study or Planned Analyses

A Statistical Analysis Plan (SAP) was not produced prior to statistical analysis as detailed in the protocol and the statistical analyses were carried out as indicated in the protocol with the exception of the following:

1. The mean profile with time curve for vital signs for each treatment is not presented.
2. The data was not summarised using the AUEC for blood pressure and pulse rate calculate using the trapezoidal rule.
3. The AUEC was not analysed using the analysis of variance with factors for subject, period and treatment

STUDY SUBJECTS

Disposition of Subjects

Twenty-four healthy male subjects were required to complete the study in its entirety. Twenty-four subjects were randomised and all of those subjects completed the study. No subjects withdrew from the study and no replacements were required.

Protocol Deviations

Three significant deviations occurred during the study as follows.

1. Post-study oral temperature was not recorded in accordance with ICH GCP, therefore the reliability of the data was not known and was not reported. All subjects were assessed by a physician prior to discharge and all were deemed to be well.
2. On May 18, 2002 (Group 2, Period 2) some blood samples for plasma concentration analysis were taken in sodium heparin blood tubes in error. The analytical laboratory carried out validation testing for use of the sodium heparin tubes and confirmed that changing the blood collection tubes from lithium heparin to sodium heparin did not alter the extraction efficiency or change in analytical methodology required.
3. Subject 101 took two single oral doses (400 mg each) of ibuprofen tablets on two consecutive days (May 18 and 19, 2002) for coryza. This was during the restriction period between Periods 2 and 3. Investigator's judgement was made and the subject was deemed eligible to continue in the study.

These protocol deviations were not considered to affect the integrity of the study.

Plasma Concentration, Pharmacokinetic, and Pharmacodynamic Evaluation

All twenty-four subjects (101 to 124) who were randomised completed the study. Subjects were considered evaluable if no more than one blood sample per period was missed. No blood samples were missed therefore all subjects were included in the data analysis.

All subjects included in the study complied with all demographic and baseline requirements. Each test treatment was administered by suitably trained study site clinical staff. No deviations to the dosing regimen were noted for any subject throughout the study.

INDIVIDUAL PLASMA CONCENTRATION DATA, PHARMACOKINETIC AND PHARMACODYNAMIC RESULTS

Analysis of Plasma Concentration Results

Plasma samples were analysed for CBD, THC and 11-hydroxy-THC according to the analytical protocol (Figure 1). Plasma concentration results are shown in tabular form (Table 2) and concentration-time graphs produced from these data (Figures 2-6).

The Lower Limit of Quantification (LLOQ) for this study was 0.1 ng/ml. The actual values measured were used when creating graphs.

Mean plasma concentrations of the relevant cannabinoids for the formulations are summarised in Table 2.

Mean plasma concentrations show that following administration of both High THC and CBD:THC formulations (Figure 2, Figure 3), THC was detectable in plasma in measurable concentrations 30-45 min after dosing, although subjects showed quite wide variability with both formulations (15-70 min). At all time points up to 180 min after dosing, mean concentrations of THC were greater following the High THC formulation (Figure 2) than CBD:THC (Figure 3). Mean 11-hydroxy-THC plasma levels (Figure 4, Figure 5) seemed generally to reflect levels of THC and were similarly greater following High THC (Figure 4) at most time points up to 180 min.

Mean plasma levels of CBD were above the level of detection about 45 min after dosing and were approximately 30-50% lower than the cor-

FIGURE 1

The following PK parameters were calculated for CBD, THC and 11-hydroxy-THC:

T_{max} Time to the maximum measured plasma concentration.

C_{max} Maximum measured plasma concentration over the time span specified.

$t_{1/2}$ Putative effective elimination half life (the initial descending portion of each plasma concentration-time graph).

AUC_{0-t} The area under the plasma concentration versus time curve, from time zero to 't' (where t = the final time of positive detection, t ≤ 24 h) as calculated by the linear trapezoidal method.

$AUC_{0-\infty}$ The area under the plasma concentration versus time curve from zero to t calculated as AUC_{0-t} plus the extrapolated amount from time t to infinity.

K_{el} Elimination rate.

TABLE 2. Mean Plasma Concentration Data

Time (min)	Analyte				
	CBD	THC		11-Hydroxy-THC	
	Test Treatment				
	CBD:THC	High THC	CBD:THC	High THC	CBD:THC
0	0.00	0.01	0.00	0.00	0.00
15	0.00	0.02	0.01	0.01	0.00
30	0.05	0.13	0.06	0.22	0.10
45	0.21	0.47	0.30	0.81	0.53
60	0.38	0.77	0.61	1.19	1.01
70	0.39	1.11	0.61	1.34	1.06
80	0.52	1.26	0.75	1.57	1.23
90	0.62	1.65	0.89	1.86	1.44
100	0.84	2.15	1.21	2.33	1.59
110	1.21	2.60	1.78	2.53	1.73
120	1.15	2.82	1.69	2.65	1.90
135	1.27	2.87	1.80	2.45	2.14
150	1.37	2.93	1.93	2.77	2.52
180	2.04	4.02	2.72	3.51	2.93
360	1.34	1.17	1.82	1.74	2.38
540	0.49	0.32	0.51	0.67	1.02
720	0.24	0.19	0.21	0.44	0.58
1440	0.00	0.03	0.01	0.16	0.16

responding levels of THC (Figure 6). Again there was quite wide variability between subjects with the time of first measurable concentration ranging from 30 to 135 min.

Following administration of High THC CBME, no subject had measurable concentrations of CBD at any time point. Following placebo, a single blood sample (60 min) from Subject 115 recorded levels of THC, CBD and 11-hydroxy-THC. Also, on the placebo dosing day, Subject 121 had measurable concentrations of THC pre-dose and at all time points post-dose. 11-Hydroxy-THC was also detected pre-dose and all time point up to 3 h post-dose. Subject 115 also had one value for THC (0.19 ng/ml at 60 min) and 11-hydroxy-THC (0.23 ng/ml at 60 min) fol-

FIGURE 2. GWPK0215 Mean Plasma THC Concentrations Following Administration of High THC

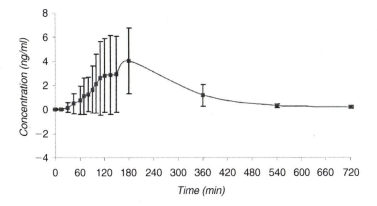

FIGURE 3. GWPK0215 Mean Plasma THC Concentrations Following Administration of CBD:THC

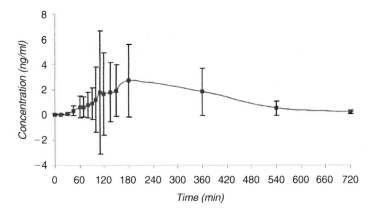

lowing administration of the placebo treatment which was above the LLOQ. The placebo treatment for subjects 115 and 121 was Period 3 and therefore followed previous High THC and CBD:THC dosing.

Analysis of Pharmacokinetic Parameters

PK parameters were calculated using WinNonlin® Professional 3.1. The model used was a non-compartmental, linear trapezoidal analysis.

FIGURE 4. GWPK0215 Mean Plasma 11-Hydroxy-THC Concentrations Following Administration of High THC

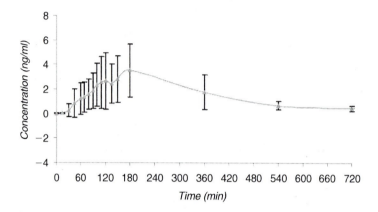

FIGURE 5. GWPK0215 Mean Plasma 11-Hydroxy-THC Concentrations Following Administration of CBD:THC

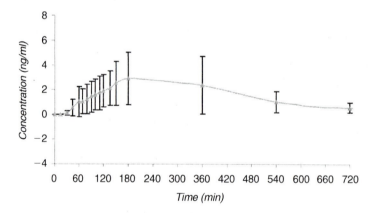

Values below the LLOQ are not considered reliable and therefore were not used when calculating PK parameters. Mean values are presented in Table 3.

Following dosing with the CBD:THC test treatment the mean C_{max}, AUC_{0-t} and $AUC_{0-\infty}$ of CBD were lower than the corresponding mean results for THC though T_{max} was similar. The $t_{1/2}$ of CBD (108.72 min) was longer than the $t_{1/2}$ of THC (84.23 min).

The PK values for each individual showed considerable inter- and

FIGURE 6. GWPK0215 Mean Plasma CBD Concentrations Following Administration of CBD:THC

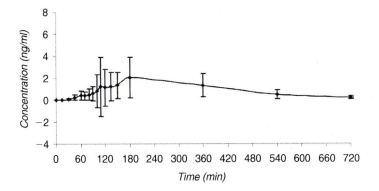

TABLE 3. Mean Pharmacokinetic Parameters

Treatment	T_{max} (min)	C_{max} (ng/ml)	$t_{1/2}$ (min)	AUC_{0-t} (min*ng/ml)	$AUC_{0-\infty}$ (min*ng/ml)
	Mean Pharmacokinetic Parameters for CBD				
CBD:THC	253	3.33	108.72	680.61	718.46
	Mean Pharmacokinetic Parameters for THC				
High THC	188	5.66	73.09	987.47	1005.90
CBD:THC	263	4.90	84.23	894.80	918.81
	Mean Pharmacokinetic Parameters for 11-Hydroxy-THC				
High THC	179	4.81	109.38	1300.47	1334.41
CBD:THC	230	4.49	130.11	1423.20	1463.67

Mean C_{max}, $t_{1/2}$, AUC_{0-t} and $AUC_{0-\infty}$ of both THC and 11-hydroxy-THC were not statistically significantly different following the High THC and CBD:THC formulations. T_{max} of both THC and 11-hydroxy-THC was later following the CBD:THC than High THC formulation though only the difference in THC T_{max} reached statistical significance (p = 0.014).

intra-subject variation in all parameters. The variability appeared to be greater for THC than for CBD. The AUC values (AUC_{0-t} and $AUC_{0-\infty}$) for THC show an approximate 8 to 10-fold difference between the lowest and highest subject values while the difference for CBD was approximately 3.5 to four-fold. Differences in C_{max} were 20 to 30-fold for THC and approximately 14-fold for CBD. Intra-subject differences in individual values for THC between treatments were smaller though differences in C_{max} of up to 5-fold and 3-fold in AUC (AUC_{0-t} and $AUC_{0-\infty}$) were observed.

Analysis of Intoxication Results

For each test treatment period, intoxication was measured using BS-11 with a score of zero indicating no intoxication and a score of 10 indicating maximum intoxication. Mean intoxication results are presented in Table 4.

On the placebo day Subject 114 scored pre-dose intoxication at a level of 5 and from 45-90 min scored a level of 6 then 5 for the remainder of the 24 h period. Two other subjects (108 and 124) scored an increased level intoxication of 1 at some time points (45-60 min and 15-60 min, respectively). Mean levels of intoxication remained low for both active test treatments throughout each 24 h post-dose period.

Following administration of High THC CBME, individual intoxication scores were below five. Seven subjects scored no intoxication at all assessment points (no score > zero). Seven subjects had at least one

TABLE 4. Mean Intoxication Following Administration of Each Test Treatment

Time (min)	Mean Intoxication		
	Placebo	High THC	CBD:THC
0	0.2	0.0	0.0
15	0.3	0.3	0.2
30	0.3	0.4	0.2
45	0.3	0.5	0.3
60	0.3	0.5	0.5
90	0.3	0.5	0.6
120	0.2	0.6	0.8
180	0.2	0.9	0.8
360	0.2	0.7	1.2
540	0.2	0.2	0.4
720	0.2	0.0	0.1
1440	0.2	0.0	0.0

Mean levels of intoxication remained below 1 throughout the 24h period following placebo dosing. Mean intoxication scores on both test treatments were very low throughout the observation period with increased levels (mean score of 1) only between 60 and 360 minutes after dosing.

score of three or greater though in four subjects this was at a single assessment point. One subject (Subject 115) recorded a score of three or greater at two consecutive time points, Subject 101 recorded scores of three between 45 and 180 min post-dose and Subject 121 recorded scores of three or greater between 30 and 180 min post-dose. The highest individual intoxication score was five (Subject 121 at 45 and 60 minutes post-dose).

Following administration of CBD:THC, nine subjects scored no intoxication at all assessment points. Ten subjects had at least one score of three or greater though in five this was at a single assessment point. Five subjects (subjects 101, 111, 112, 113 and 116) recorded a score of three or greater at two consecutive time points. The highest individual intoxication score was 10 (Subject 112 at a single time point post-dose).

Recorded intoxication scores do not seem to show a direct relationship to plasma concentrations of THC and/or 11-hydroxy-THC either within or between subjects. The times of intoxication scores in individual subjects do not seem to relate consistently with the timing of increases in plasma concentrations or maximal concentrations of THC or 11-hydroxy-THC. Neither is there an apparent relationship between subjects reporting intoxication and those with the highest plasma levels of THC or 11-hydroxy-THC. The maximum intoxication score of 10 reported by Subject 112 occurred 360 minutes post-administration of CBD:THC. This maximal intoxication score was not associated with any report of AEs typical of intoxication (e.g., somnolence, dizziness). Vital signs at this time were only a little changed from pre-dose–pulse 68 (-4), systolic BP 106 (-16) diastolic BP 63 (-4) and do not suggest significant cannabinoid effects. However, the score of 10 coincided with a substantial increase in plasma levels of both THC (3.56 ng/ml) and 11-hydroxy-THC (3.96 ng/ml) compared with both the previous (0.21 and 0.48 ng/ml, respectively) and subsequent measurements (0.77 and 1.88 ng/ml, respectively) at which much lower intoxication scores were reported (0 and 3, respectively). On the day that the High THC was administered the highest intoxication score recorded by this subject was three at 6 h post-dose even though during this dosing period higher plasma levels of THC (2.45 ng/ml) were recorded compared with the CBD:THC test treatment. Plasma levels of 11-hydroxy-THC were a little lower on this occasion.

Analysis of Safety Parameters

For each of the BP and pulse parameters descriptive statistics (n, mean, SD, median, minimum and maximum) were presented at each time point by test treatment. In addition, the calculations were performed for the absolute change from pre-dose. Mean values and mean changes from baseline were similar across all treatments.

Descriptive statistics (n, mean, SD, median, minimum and maximum) were recorded for the ECG parameters (heart rate, PR interval, QT interval and QRS width) pre-dose and at each time point by test treatment. ECG intervals were expertly read by Cardio Analytics for each of the parameters above. There were no notable changes in the ECG parameters.

Eight subjects (33%) rated the placebo test treatment as very unpleasant or unpleasant compared with 18 subjects (75%) for both the THC and CBD:THC treatments.

One subject (4%) thought the placebo treatment had an unpleasant smell compared with four (17%) subjects who thought the High THC treatment smelt unpleasant. Four (17%) subjects thought the CBD:THC smelt unpleasant and two (8%) very unpleasant. Eleven subjects (46%) for each treatment reported that they were unaware of the smell.

All three test treatments resulted in increased saliva produced with 13 subjects (54%) reporting more saliva following administration of placebo, 16 subjects (66%) with High THC and 17 subjects (71%) with CBD:THC.

The majority of subjects reported that they thought all or most of the test treatments were absorbed in the mouth. Only six subjects (25%) after placebo and High THC thought that some was swallowed and four subjects (17%) after CBD:THC reported some was swallowed.

Most subjects reported no other effects or sensations following administration of each test treatment. Four subjects (17%) reported other effects following administration of placebo, nine subjects (38%) following administration of High THC and 10 subjects (42%) following administration of CBD:THC.

The study was carried out in healthy subjects, none of whom were not taking a regular course of any other medication.

Plasma Concentration Conclusions

Mean plasma concentrations show that following administration of both High THC and CBD:THC formulations, CBD and/or THC were detectable in plasma in measurable concentrations 30-45 min after dosing, although individual subjects showed quite wide variability, 15 to 135 min, to appearance of measurable concentrations. At all time points up to 180 min after dosing mean concentrations of THC were greater following the High THC formulation than CBD:THC. Concentrations of THC were also greater than corresponding concentrations of CBD following the CBD:THC treatment.

There was considerable individual variability in peak plasma concentrations (C_{max}) of both CBMEs. THC C_{max} ranged from 0.69 ng/ml to 14.2 ng/ml and from 0.75 ng/ml to 24.63 ng/ml for the High THC and CBD:THC formulations, respectively. CBD C_{max} following the CBD:THC formulation ranged from 0.96 ng/ml to13.64 ng/ml.

Following administration of High THC CBME, no subject had measurable concentrations of CBD at any time point. Following placebo, a single blood sample (60 min) from Subject 115 had recorded measurable levels of THC, CBD and 11-hydroxy-THC. This sample was re-analysed by the analytical laboratory, however the result may be due to an analytical anomaly. Also on the placebo dosing day Subject 121 had measurable concentrations of THC pre-dose and at all time points post-dose and 11-hydroxy-THC was also measured pre-dose and all time point up to 3 hours post-dose. The placebo treatment in this subject was Period 3 and therefore followed previous High THC and CBD:THC dosing. As there was no carryover from Period 1 to Period 2 in this subject it is unclear whether the THC detected on the placebo day is due to carryover from the previous treatment or a protocol violation in respect of abstention from cannabis.

Pharmacokinetic Conclusions

There were no statistically significant differences in mean C_{max}, $t_{1/2}$, AUC_{0-t} and $AUC_{0-\infty}$ of both THC and 11-hydroxy-THC between the High THC and CBD:THC formulations. THC T_{max} was statistically significantly later following CBD:THC than High THC ($p = 0.014$) and this was the only statistically significant difference in PK parameters between the treatments. Following the CBD:THC formulation the C_{max} and AUC of CBD were lower than the corresponding results for THC and the $t_{1/2}$ of CBD (108.72 min) was longer than the $t_{1/2}$ of THC (84.23 min). The PK values for each individual show considerable inter- and intra-subject variation in all parameters. The variability appears to be greater for THC than for CBD. The AUC values (AUC_{0-t} and $AUC_{0-\infty}$) for THC show an approximate 8 to 10-fold difference between the lowest and highest subject values while the difference for CBD was approximately 3.5 to 4-fold. Differences in C_{max} were 20 to 30 fold for THC and approximately 14-fold for CBD. Intra-subject differences in values for THC between treatments were smaller though differences in C_{max} of up to 5-fold and 3-fold in AUC (AUC_{0-t} and $AUC_{0-\infty}$) were observed.

Other than a single isolated significant difference in T_{max} there were no significant differences in PK parameters between the CBD:THC and High THC formulations. It is unclear whether this significant difference reflects a true or spurious difference in the rates of absorption from the formulations, however, the difference is small and unlikely to be of clinical significance considering the high level of inter- and intra-subject variability in PK. The bioavailability of THC appears to be greater than that of CBD.

Intoxication Conclusions

Mean intoxication scores on both CBME treatments were very low throughout the observation period. The majority of subjects scored zero for the majority of assessment points and there were few scores greater than three on the 11 box scale. One subject recorded a maximal score of 10 at a single (6 h) assessment point following CBD:THC. No AEs were reported and vital signs showed only a slight change from pre-dose at this time, therefore it is uncertain that this reflects an accurate assessment. Recorded intoxication scores do not seem to show a direct relationship to plasma concentrations of THC and/or 11-hydroxy-THC either within or between subjects. The time of intoxication scores in individual subjects do not seem to relate consistently with the timing of increases in plasma concentrations or maximal concentrations of THC or 11-hydroxy-THC. Neither is there an apparent relationship between subjects reporting intoxication and those with the highest plasma levels of THC or 11-hydroxy-THC.

Palatability Conclusions

Both active test treatments, but not placebo, were considered by the majority of the subjects to have an unpleasant or very unpleasant taste. Therefore it can be concluded that the THC and/or CBD, and not the excipients, result in an increased incidence of unpleasant taste. The majority of subjects reported that they were not aware of a smell from the test treatment or that they thought it smelt neither pleasant or unpleasant. Therefore it can be concluded that for the majority of subjects THC and/or CBD in the test treatments used in this study do not have an unpleasant smell. All three test treatments were reported to have increased saliva with a marginally higher incidence from the CBME containing treatments. Most subjects perceived that all or most of the test treatments were absorbed in the mouth.

Other Effects or Sensations

The incidence of other effects or sensations following administration of each test treatment was greater for the CBME treatments than for placebo though the majority of subjects on all treatments reported no such effects.

ADVERSE EVENTS (AES)

Brief Summary of Adverse Events

During the study 87 AEs were recorded in 20 subjects (Table 5), and of these, 78 were considered to be related to the test treatment (Table 6). Following the administration of placebo, 5 subjects experienced treatment emergent treatment related AEs (Table 6). Following administration of the THC test treatment 16 subjects (66%) experienced treatment emergent treatment related AEs and 18 subjects (75%) experienced treatment emergent treatment related AEs following administration of CBD:THC (Table 6). All the AEs experienced were classified as mild and only one event (Subject 101, coryza) required treatment with medication. None of the subjects withdrew due to AEs. One AE in Subject 107 (left shoulder muscular strain) was lost to follow up.

The most common treatment emergent treatment related AE experienced was throat irritation (six subjects following administration of High THC and eight subjects following administration of CBD:THC), which was not experienced in the subjects during placebo treatment. Dizziness was the second most commonly experienced treatment emergent treatment related AE following the administration of High THC (six subjects). This was followed by somnolence, oral paraesthesia and headache.

Analysis of Adverse Events

Table 7 summarises the number of subjects who reported treatment emergent treatment related AEs by System Organ Class (SOC). There

TABLE 5. Summary of Adverse Events–Treatment Emergent All Causality

Event	Placebo	High THC	CBD:THC
No. of subjects with ≥ 1 event	6 (25.0%)	16 (66.7%)	18 (75.0%)
Eye disorders	0	1	0
Vision blurred		1	
Gastrointestinal disorders	1	9	14
Diarrhoea NOS			1
Glossitis	1		1
Nausea		2	2
Oral discomfort		1	1
Oral pain		1	1
Throat irritation		6	8
Tongue oedema			1
Vomiting NOS			1
General disorders and administration site conditions	2	2	1
Feeling of relaxation	2	1	
Lethargy	2	1	1
Injury, poisoning and procedural complications	1	1	2
Drug toxicity NOS		1	2
Splinter	1		
Musculoskeletal and connective tissue disorders	0	1	2
Muscle strain			1
Muscle twitching			1
Rib fracture		1	
Nervous system disorders	4	11	10
Burning sensation NOS		1	1
Dizziness	1	6	2
Dysgeusia			2
Headache NOS	2		3
Paraesthesia		1	
Paraesthesia oral NOS	1	2	3
Somnolence		4	3
Vasovagal attack (LLT Syncope vasovagal)		1	
Respiratory, thoracic and mediastinal disorders	0	3	2
Cough		1	
Rhinitis NOS		2	2
Skin and subcutaneous tissue disorders	1	0	0
Localised skin reaction	1		
Vascular disorders	0	1	0
Hot flushes NOS		1	

TABLE 6. Summary of Adverse Events–Treatment Emergent Treatment Related

Event	Placebo	High THC	CBD:THC
No. of subjects with ≥ *1 event*	5 (20.8%)	16 (66.7%)	18 (75.0%)
Eye disorders	0	1	0
Vision blurred		1	
Gastrointestinal disorders	1	9	14
Diarrhoea NOS			1
Glossitis	1		1
Nausea		2	2
Oral discomfort		1	1
Oral pain		1	1
Throat irritation		6	8
Tongue oedema			1
Vomiting NOS			1
General disorders and administration site conditions	2	2	1
Feeling of relaxation	2	1	
Lethargy	2	1	1
Injury, poisoning and procedural complications	0	1	2
Drug toxicity NOS		1	2
Musculoskeletal and connective tissue disorders	0	0	1
Muscle twitching			1
Nervous system disorders	4	10	10
Burning sensation NOS		1	1
Dizziness	1	6	2
Dysgeusia			2
Headache NOS	2		3
Paraesthesia		1	
Paraesthesia oral NOS	1	2	3
Somnolence		3	3
Vasovagal attack (LLT syncope vasovagal)		1	
Respiratory, thoracic and mediastinal disorders	0	1	1
Cough		1	
Rhinitis NOS			1
Vascular disorders	0	1	0
Hot flushes NOS		1	

Note: treatment related = definitely, probably, possibly related

TABLE 7. Summary of Number of Subjects Who Experienced at Least One AE per SOC–Treatment Emergent Treatment Related

Event	Placebo (n = 24)	High THC (n = 24)	CBD:THC (n = 24)
No. of subjects with ≥ 1 event	5 (20.8%)	16 (66.7%)	18 (75.0%)
Eye disorders	0	1	0
Gastrointestinal disorders	1	9	14
General disorders and administration site conditions	2	2	1
Injury, poisoning and procedural complications	0	1	2
Musculoskeletal and connective tissue disorders	0	0	1
Nervous system disorders	4	10	10
Respiratory, thoracic and mediastinal disorders	0	1	1
Vascular disorders	0	1	0

were no deaths or serious AEs during the study, and no withdrawals attributed to AEs.

Clinical Laboratory Evaluation

All out of range values noted were considered by the Principal Investigator to be "not clinically significant." There were no clinically significant laboratory findings during the study. Several pre- and post-study results were out of the normal range but were not considered clinically significant. There were no statistically significant changes from pre- to post-study in any of the laboratory parameters. There were no notable changes, patterns or trends within the values from pre- and post-study in individual subjects.

Vital Signs, Physical Findings and Other Observations Related to Safety

There were no notable changes in diastolic BP during the study. There was a small transient increase in the mean pulse rate after 15 min during the High THC and CBD:THC periods. After three hours the mean systolic BP decreased by 10.3 mmHg during the High THC period, by 4.4 mmHg in the CBD:THC period, and 5.1 mmHg during the placebo period. After 12 hours the mean pulse, systolic and diastolic BP values were close to the pre-dose values for all treatments.

No clinically significant changes in physical examination findings were noted from pre- to post-study. Only one change was noted in one subject, which began pre-dose and was not considered to be related to the test treatment. Each subject was asked about their previous medical history at screening. No events were considered to significant in relation to this study. There was no notable trend or pattern in the HR (BPM), PR Interval (msecs), QT_c (msecs), QRS width (msecs) in comparison to placebo. Two subjects had a borderline QT_c after dosing compared to pre-dose values. Subject 115 (CBD:THC period) had an increased QT_c of 41 msec (borderline) after 2 hours, this returned to normal after 12 h. Subject 119 (placebo period) had an increased QT_c of 35 msec (borderline) after 1 h and 33 msec after 12 h. In the opinion of the Investigator both borderline QT_c increases from pre-dose were considered not clinically significant. The ECGs taken during the study were read manually.

Safety Conclusions

The results of this study show that all three test treatments were well tolerated. CBD:THC had the most AEs followed by the THC group and then the placebo group. High THC and CBD:THC had a greater number of subjects who experienced intoxication type AEs and application site type reactions than placebo. The most common overall AE experienced was throat irritation, followed by dizziness, somnolence, oral paraesthesia and then headache. All the events were mild, one required treatment and one event was lost to follow-up.

DISCUSSION AND OVERALL CONCLUSIONS

All three test treatments administered in the study were well tolerated by all subjects. There were no AEs which resulted in any subject withdrawals from the study. Intoxication scores in the study were similarly low for both active treatments and did not appear to be directly related to plasma concentrations of THC and/or 11-hydroxy-THC and intoxication. There were no statistically significant differences in mean C_{max}, $t_{1/2}$, AUC_{0-t} and $AUC_{0-\infty}$ of both THC and 11-hydroxy-THC between the High THC and CBD:THC formulations. THC T_{max} was statistically significantly later (262.7 mins compared with 187.7 mins) following

CBD:THC than High THC (p = 0.014) and this was the only statistically significant difference in PK parameters between the treatments. It is possible that the presence of CBD in the CBD:THC formulation delays the absorption of THC.

There was wide inter- and intra-subject variability in PK parameters with up to 10-fold differences in THC AUC between subjects and even greater differences in C_{max}. Results suggest that there are no overall statistically significant differences between the PK parameters of High THC and CBD:THC other than a delay in T_{max}. Considering the wide inter- and intra-subject variability in PK parameters, including T_{max}, this is unlikely to be clinically important in a medication that is self-titrated by the patient.

REFERENCES

Agurell, S., M. Halldin, J. E. Lindgren, A. Ohlsson, M. Widman, H. Gillespie, and L. Hollister. 1986. Pharmacokinetics and metabolism of delta 1-tetrahydrocannabinol and other cannabinoids with emphasis on man. *Pharmacol Rev* 38 (1):21-43.

Association, British Medical. 1997. *Therapeutic uses of cannabis.* Amsterdam: Harwood Academic Publishers.

Baker, P. B., T. A. Gough, and B. J. Taylor. 1983. The physical and chemical features of Cannabis plants grown in the United Kingdom of Great Britain and Northern Ireland from seeds of known origin–Part II: Second generation studies. *Bull Narc* 35 (1):51-62.

Lords, House of. 1998. *Cannabis: The scientific and medical evidence.* London: House of Lords Select Committee on Science and Technology, Stationery Office.

Maykut, M. O. 1985. Health consequences of acute and chronic marihuana use. *Prog Neuropsychopharmacol Biol Psychiatry* 9 (3):209-38.

McPartland, J. M., and E. B. Russo. 2001. Cannabis and cannabis extracts: Greater than the sum of their parts? *J Cannabis Therap* 1(3-4):103-132.

Pharmaceuticals, GW. 2002. Investigators brochure for cannabis based medicine extract (CBME). Porton Down: GW Pharmaceuticals.

Portenoy, R. K. 1990. Chronic opioid therapy in nonmalignant pain. *J Pain Symptom Manage* 5 (1 Suppl):S46-62.

Porter, J., and H. Jick. 1980. Addiction rare in patients treated with narcotics. *N Engl J Med* 302 (2):123.

Whittle, B. A., G. W. Guy, and P. Robson. 2001. Prospects for new cannabis-based prescription medicines. *J Cannabis Therap* 1(3-4):183-205.

Cannabis
and Cannabis Based Medicine Extracts:
Additional Results

Ethan Russo

SUMMARY. This study reviews results in recent human clinical trials with cannabis based medicine extract (CBME), THC or cannabis.

In a study performed at Queen's Square, London, both High THC and THC:CBD fixed ratio sublingual CBME demonstrated significant benefits on mean maximum cystometric capacity, mean daytime frequency of urination, frequency of nocturia, and mean daily episodes of incontinence in 11 multiple sclerosis patients with intractable lower urinary tract symptoms.

A Phase II clinical study in Oxford, England with 24 MS and intractable pain patients was performed as a consecutive series of double-blind, randomized, placebo-controlled single patient cross-over trials with sublingual CBME. Pain scores on visual analogue scales were significantly improved over placebo with both High THC and High CBD CBME. Subjectively, spasm was significantly improved with High THC and THC:CBD fixed ratio extracts. Spasticity was also subjectively improved with the High THC CBME. All three extracts significantly improved objective measures of spasticity, while the High THC and

Ethan Russo, MD, is a Clinical Child and Adult Neurologist, Clinical Assistant Professor of Medicine, University of Washington, and Adjunct Associate Professor of Pharmacy, University of Montana, 2235 Wylie Avenue, Missoula, MT 59802 USA (E-mail: erusso@montanadsl.net).

[Haworth co-indexing entry note]: "Cannabis and Cannabis Based Medicine Extracts: Additional Results." Russo, Ethan. Co-published simultaneously in *Journal of Cannabis Therapeutics* (The Haworth Integrative Healing Press, an imprint of The Haworth Press, Inc.) Vol. 3, No. 4, 2003, pp. 153-161; and: *Cannabis: From Pariah to Prescription* (ed: Ethan Russo) The Haworth Integrative Healing Press, an imprint of The Haworth Press, Inc., 2003, pp. 153-161. Single or multiple copies of this article are available for a fee from The Haworth Document Delivery Service [1-800-HAWORTH, 9:00 a.m. - 5:00 p.m. (EST). E-mail address: docdelivery@haworthpress.com].

http://www.haworthpress.com/store/product.asp?sku=J175
© 2003 by The Haworth Press, Inc. All rights reserved.
10.1300/J175v03n04_03

THC:CBD fixed ratio CBME significantly improved objective measures of spasm.

In 34 intractable pain patients in Great Yarmouth, England, seven experienced substantial improvement over best available conventional treatment with CBME, 13 moderate, and eight some benefit. Many extended the range of their activities of daily living with acceptable levels of adverse effects.

Preliminary results of four Phase III clinical trials of CBME by GW Pharmaceuticals have revealed highly significant benefits in neuropathic pain in MS, pain and sleep disturbance in MS and other neurological diseases, multiple symptoms in MS, and neuropathic pain in brachial plexus injury, respectively. Most patients attained good symptomatic control with minimal side effects.

In Germany, a recent Phase II clinical trial has demonstrated significant benefit of oral THC in treatment of the tics of Tourette syndrome. *[Article copies available for a fee from The Haworth Document Delivery Service: 1-800-HAWORTH. E-mail address: <docdelivery@haworthpress.com> Website: <http://www.HaworthPress.com> © 2003 by The Haworth Press, Inc. All rights reserved.]*

KEYWORDS. Medical marijuana, cannabis, alternative delivery systems, THC, cannabidiol, CBD, multiple sclerosis, chronic pain, Tourette syndrome, brachial plexus injury, pharmacotherapy

In 2001, interim results of a study of cannabis based medicine extracts (CBMEs) in bladder dysfunction were presented at the meeting of the International Association for Cannabis as Medicine (IACM) (Brady et al. 2001). A high-THC CBME and 1:1 THC:CBD CBME were compared to placebo in 17 multiple sclerosis patients with refractory lower urinary tract symptoms (LUTS). Eleven patients had evaluable data. Doses of up to 10 mg THC or 10 mg of THC and 10 mg of CBD were utilized. Mean maximum cystometric capacity (MCC) increased from 287 ml at baseline to 344 ml after eight weeks of CBME treatment (with 24 h of no drug). After 16 weeks, the bladder capacity measured 425 ml at maximum THC:CBD dosage. Mean daytime frequency or urination went from 9.3 to 7.5 with CBD:THC 1:1 and 6.9 with high-THC CBME. Similarly, nocturia episodes fell from 2.7 at baseline to 1.4 with the 1:1 mixture, and 1.5 with high-THC. Additionally, mean episodes of daily incontinence fell from a baseline of 2.1 to 1.0 with CBD:THC and 0.7 with high-THC CBME. These results will soon be published more formally.

In the past year, a small clinical trial of THC and a cannabis extract was performed with 16 subjects. Neither was observed to reduce spasticity, and adverse events were reported in the extract group (Killestein et al. 2002). Numerous criticisms were subsequently voiced in this regard (Russo 2003). Among these were that the plant extract was poorly categorized; in fact, it contained a fixed of THC to CBD with maximum doses of 5 mg of THC and 2 mg of CBD per day. The study additionally employed oral administration with no real dose titration. An additional study in Switzerland with more patients (57) and doses of up to 15 mg THC with 6 mg CBD divided tid has provided better results with reduction in spasms to the $p < 0.05$ level and no significant side effects vs. placebo (Vaney et al. 2002). A study of an even larger cohort in the UK is pending publication.

The results of a Phase II study of CBME have recently been published (Wade et al. 2003). This clinical trial was performed in Oxford, England with 24 subjects with treatment-resistant MS, spinal or brachial plexus injury comparing THC, CBD, THC:CBD, and placebo sublingual extracts employing consecutive series of double-blind, randomized, placebo-controlled single patient cross-over trials. Subjective and objective measures of pain, spasticity, spasm et al. were monitored along with adverse effects. Results were monitored employing subjective and objective blinded ratings and visual analogue scales (VAS). Twenty of the subjects completed the trial. Results with statistical significance included:

1. Pain scores were improved with both high-THC and high-CBD CBME vs. placebo ($p < 0.05$) (Figure 1).
2. Spasm was improved with both the high-THC and fixed-ratio THC:CBD CBME ($p < 0.05$) (Figure 2).
3. Similarly, spasticity was improved subjectively with the high-THC preparation ($p < 0.05$) (Figure 3).
4. As might be surmised, the high-THC CBME improved subjective measure of appetite ($p < 0.05$) (Figure 4).
5. The fixed-ratio THC:CBD CBME produced the best improvement in subjective sleep ($p < 0.05$) (Figure 5).
6. Turning to blinded objective measures, all three extracts, high-THC, high-CBD and fixed-ratio THC:CBD CBME improved spasticity on a numerical symptom scale ($p < 0.05$) (Figure 6).
7. Similarly, the high-THC and THC:CBD fixed-ratio CBME's yielded statistically significant objective improvement in spasm frequency ($p < 0.05$) (Figure 7).

FIGURE 1. Pain Improvement, N = 20, Daily VAS

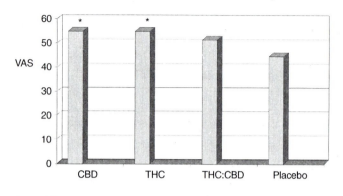

* = p < 0.05
Adapted from Wade DT, Robson P et al. 2003, Clin Rehab 17:18-26.

FIGURE 2. Spasm, N = 20, Daily VAS

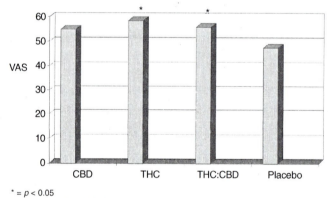

* = p < 0.05
Adapted from Wade DT, Robson P et al. 2003, Clin Rehab 17:18-26.

Adverse effects in the trial were predictable and well tolerated.

Additional Phase II work has been pursued in chronic pain patients intractable to conventional pharmacotherapy by the team of Notcutt et al. at James Paget Hospital in Great Yarmouth, UK. This work is pending more formal publication, but has been reported in 9 abstracts in the *Journal of Cannabis Therapeutics* from the 2001 meeting of the International Association for Cannabis as Medicine in Berlin (Notcutt 2002; Notcutt et al. 2002, 2002, 2002, 2002, 2002, 2002, 2002, 2002), as well

FIGURE 3. Spasticity, N = 20, Daily VAS

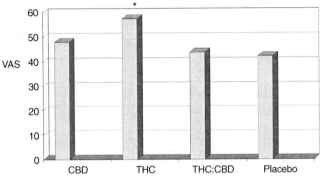

* = p < 0.05
Adapted from Wade DT, Robson P et al. 2003, Clin Rehab 17:18-26.

FIGURE 4. Appetite, N = 20, Daily VAS

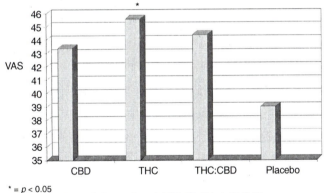

* = p < 0.05
Adapted from Wade DT, Robson P et al. 2003, Clin Rehab 17:18-26.

as the 2002 meeting of the International Cannabinoid Research Society in Asilomar, California (Notcutt 2003). Briefly stated, 34 N-of-1 studies were performed in a cohort of inadequately controlled pain patients, including those with MS (16), chronic back pain and sciatica (eight), other neuropathic pain (five), complex regional pain syndrome (CRPS, or "reflex sympathetic dystrophy") (two), and polyarthralgia, stiff man syndrome and myopathy (one each). Subjects included both cannabis-experienced and cannabis-naïve individuals. After a two-week base-

FIGURE 5. Sleep, N = 20, Daily VAS

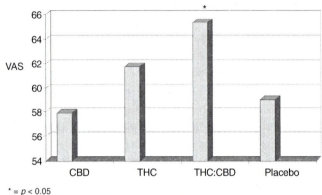

* = $p < 0.05$
Adapted from Wade DT, Robson P et al. 2003, Clin Rehab 17:18-26.

FIGURE 6. Spasticity Severity, Numerical Symptom Scale, N = 20, Observer Rated

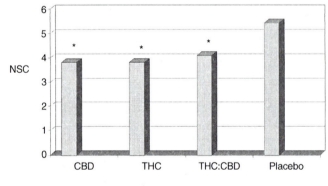

* = $p < 0.05$
Adapted from Wade DT, Robson P et al. 2003, Clin Rehab 17:18-26.

line evaluation, a subsequent two-week open-label titration trial (one spray every 30 minutes to a limit of four with subsequent patient-directed upward titration) was pursued with fixed-ratio THC:CBD, followed by two separate four-week double-blind randomized trials of one week each of high-THC, high-CBD, fixed-ratio THC:CBD or placebo. General benefits were noted in CBME groups in pain, sleep, depression, activity and overall health compared to placebo. Interestingly, individ-

FIGURE 7. Spasm Frequency, Numerical Symptom Scale, N = 20, Observer Rated

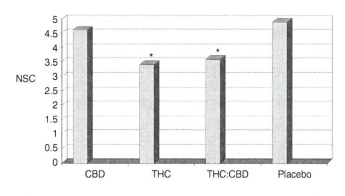

* = *p* < 0.05
Adapted from Wade DT, Robson P et al. 2003, Clin Rehab 17:18-26.

ual dose requirements varied tremendously in the cohort, with symptomatic control requiring 5-80 mg per day of THC, CBD or the mixture. Seven patients experienced substantial improvement with CBME over best available conventional treatment, while 13 (32.8%) had moderate benefit, eight (23.5%) had "some" benefit, and six (17.6%) had none. Some dysphoria occurred at dose initiation, particularly in cannabis-naïve patients, but passed in 2-3 hours. Postural hypotension occurred in three patients with dose overload, while lesser adverse effects included mucosal stinging, staining of teeth, taste change and dry skin. Randomization was broken in four patients, one was removed due to distress, one continued single-blind after marital issues, one continued after an orthostatic hypotension event, and one continued single-blind after a gastroenteritis, deemed unrelated. Overall, the CBME was felt to be effective and acceptable to patients. Twenty-nine patients (85%) elected to continue into a long-term safety study. In the aftermath of this study, subjects were noted to be able to engage in many high level pursuits of which they were previously incapable.

In November 2002, preliminary results from four Phase III randomized, double-blind, placebo controlled Phase III clinical trials in the UK with 350 patients were released by GW Pharmaceuticals, and are available online: <http://www.gwpharm.com/news_pres_05_nov_02.html>. Results from these studies included highly statistically significant reductions in neuropathic pain, spasticity and sleep disturbance. The topics of the studies included the following:

1. Neuropathic pain in MS
2. Pain and sleep disturbance in MS and other neurological conditions
3. Multiple symptoms in MS
4. Neuropathic pain in brachial plexus injury

In the Phase III study of neuropathic pain in multiple sclerosis, 66 patients were studied in double-blind parallel groups with THC:CBD vs. placebo. Pain relief with THC:CBD CBME was greater than placebo ($p < 0.01$), and sleep disturbance was relieved to the same level ($p <$ 0.01).

In the Phase III chronic refractory pain trial, 70 subjects with MS and other conditions were examined in double-blind parallel groups with THC:CBD CBME. Pain relief was observed with decreased usage of rescue medication as compared to placebo ($p < .05$), and sleep disturbance was also diminished ($p < .05$).

A larger cohort of 160 MS patients was studied in a third double-blind parallel group examining the fixed-ratio THC:CBD CBME. Spasticity was improved to a highly statistically significant degree ($p <$ 0.01), while trends of improvement were also noted for a variety of other associated symptoms.

Finally, a fourth study examined brachial plexus injury, an intractable pain syndrome most often encountered after motorcycle accidents in the UK. In the largest study and first ever controlled clinical trial in this disorder, 48 subjects were studied in a double-blind crossover protocol comparing THC, THC:CBD and placebo. THC and THC:CBD CBME both reduced pain greater than placebo to a highly statistically significant degree ($p < 0.01$). THC and THC:CBD CBME both reduced sleep disturbance to a significant degree ($p < 0.05$).

Certain other features of the trials deserve emphasis. Firstly, after 350 patient-years of experience with CBME, the improvements in clinical parameters involved were attained above and beyond those achievable with best-available "conventional" pharmaceuticals. Additionally, with self-titration, most patients were capable of alleviating their symptomatology without adverse effects on activities of daily living (ADL). The safety profile was judged, "excellent."

At the time of this writing (May 2003), five additional Phase III clinical trials including cancer pain and spinal cord injury are in process, and will be completed in 2003, at which time a cumulative 1000 patients shall have been studied.

Finally, a team in Germany has recently published a Phase II study of oral THC in Tourette syndrome (TS) (Muller-Vahl, Schneider et al. 2003), in which 24 patients were treated over 6 weeks with up to 10 mg a day in a randomized, double-blind, placebo-controlled study. Tics were assessed by a variety of measures both subjectively and objectively. Seven patients dropped out, but only one due to adverse effects. Significant benefits were noted ($p < 0.05$) in a variety of measures with no serious adverse effects. The authors concluded that THC was safe and effective in treatment of tics associated with Tourette syndrome.

REFERENCES

Brady, C.M., R. DasGupta, O.J. Wiseman, K.J. Berkley, and C.J. Fowler. 2001. Acute and chronic effects of cannabis based medicinal extract on refractory lower urinary tract dysfunction in patients with advanced multiple sclerosis-Early results. Paper read at Congress of the International Association for Cannabis as Medicine, October 26, at Berlin, Germany.

Muller-Vahl, K.R., U. Schneider, H. Prevedel, K. Theloe, H. Kolbe, T. Daldrup, and H.M. Emrich. 2003. Delta9-tetrahydrocannabinol (THC) is effective in the treatment of tics in Tourette syndrome: A 6-week randomized trial. *J Clin Psychiatry* 64 (4):459-465.

Notcutt, W. 2002. Medicinal cannabis extracts in chronic pain. *J Cannabis Therapeutics* 2 (2):101-102.

_____2003. Cannabis in the treatment of neuropathic pain. In *Medicinal uses of cannabis and cannabinoids.*, edited by B.A. Whittle, G.W. Guy and P. Robson. London: Pharmaceutical Press.

Notcutt, W., M. Price, R. Miller, S. Newport, C. Sansom, and S. Simmonds. 2002. Medicinal cannabis extract in chronic pain: (1) Design of a comparative "N of 1" primary study (CBME-1). *J Cannabis Therapeutics* 2 (2):93-94.

_____2002. Medicinal cannabis extract in chronic pain: (6) Design of long term safety study (CBME-SAFEX). *J Cannabis Therapeutics* 2 (2):98-99.

_____2002. Medicinal cannabis extract in chronic pain: (7) Results from long term safety extension study (CBME-SAFEX). *J Cannabis Therapeutics* 2 (2):99-100.

_____2002. Medicinal cannabis extract in chronic pain: (8) Evaluation of THC-CBD against THC in the management of chronic pain. *J Cannabis Therapeutics* 2 (2):100-101.

_____2002. Medicinal cannabis extracts in chronic pain: (2) Comparison of two patients with back pain and sciatica. *J Cannabis Therapeutics* 2 (2):94-95.

_____2002. Medicinal cannabis extracts in chronic pain: (3) Comparison of two patients with multiple sclerosis. *J Cannabis Therapeutics* 2 (2):95-96.

_____2002. Medicinal cannabis extracts in chronic pain: (4) Cannabidiol modification of psycho-active effects of delta-9-THC. *J Cannabis Therapeutics* 2 (2):96-97.

_____2002. Medicinal cannabis extracts in chronic pain: (5) Cognitive function and blood cannabinoid levels. *J Cannabis Therapeutics* 2 (2):97-98.

Wade, D.T., P. Robson, H. House, P. Makela, and J. Aram. 2003. A preliminary controlled study to determine whether whole-plant cannabis extracts can improve intractable neurogenic symptoms. *Clin Rehab* 17:18-26.

Future of Cannabis and Cannabinoids in Therapeutics

Ethan Russo

SUMMARY. This study reviews human clinical experience to date with several synthetic cannabinoids, including nabilone, levonantradol, ajulemic acid (CT3), dexanabinol (HU-211), HU-308, and SR141716 (Rimonabant®). Additionally, the concept of "clinical endogenous cannabinoid deficiency" is explored as a possible factor in migraine, idiopathic bowel disease, fibromyalgia and other clinical pain states. The concept of analgesic synergy of cannabinoids and opioids is addressed. A cannabinoid-mediated improvement in night vision at the retinal level is discussed, as well as its potential application to treatment of retinitis pigmentosa and other conditions. Additionally noted is the role of cannabinoid treatment in neuroprotection and its application to closed head injury, cerebrovascular accidents, and CNS degenerative diseases including Alzheimer, Huntington, Parkinson diseases and ALS.

Excellent clinical results employing cannabis based medicine extracts (CBME) in spasticity and spasms of MS suggests extension of such treatment to other spasmodic and dystonic conditions.

Finally, controversial areas of cannabinoid treatment in obstetrics, gynecology and pediatrics are addressed along with a rationale for such interventions. *[Article copies available for a fee from The Haworth Document Delivery Service: 1-800-HAWORTH. E-mail address: <docdelivery@haworthpress.com> Website: <http://www.HaworthPress.com> © 2003 by The Haworth Press, Inc. All rights reserved.]*

Ethan Russo, MD, is a Clinical Child and Adult Neurologist, Clinical Assistant Professor of Medicine, University of Washington, and Adjunct Associate Professor of Pharmacy, University of Montana, 2235 Wylie Avenue, Missoula, MT 59802 USA (E-mail: erusso@montanadsl.net).

[Haworth co-indexing entry note]: "Future of Cannabis and Cannabinoids in Therapeutics." Russo, Ethan. Co-published simultaneously in *Journal of Cannabis Therapeutics* (The Haworth Integrative Healing Press, an imprint of The Haworth Press, Inc.) Vol. 3, No. 4, 2003, pp. 163-174; and: *Cannabis: From Pariah to Prescription* (ed: Ethan Russo) The Haworth Integrative Healing Press, an imprint of The Haworth Press, Inc., 2003, pp. 163-174. Single or multiple copies of this article are available for a fee from The Haworth Document Delivery Service [1-800-HAWORTH, 9:00 a.m. - 5:00 p.m. (EST). E-mail address: docdelivery@haworthpress.com].

http://www.haworthpress.com/store/product.asp?sku=J175
© 2003 by The Haworth Press, Inc. All rights reserved.
10.1300/J175v03n04_04

KEYWORDS. Nabilone, levonantradol, ajulemic acid, dexanabinol, HU-308, SR141716, cannabis, medical marijuana, migraine, idiopathic bowel disease, fibromyalgia, night vision, retinitis pigmentosa, neuroprotection, dystonia, obstetrics, gynecology, pediatrics

INTRODUCTION

As is evident from preceding information in this publication, an increasingly bright future seems to be on the horizon for cannabis therapeutics, whether herbally-based or designed to utilize its various components. The pros and cons of cannabis proper, whether smoked, ingested orally, or vaporized have been previously addressed. A wide variety of delivery systems is possible in the future. The present selection will detail additional preparations, particularly synthetic cannabinoids, and discuss how they and cannabis-based pharmaceuticals may be applied in future clinical therapeutics.

NABILONE

Nabilone is a synthetic cannabinoid, pharmacologically similar to THC, but with higher potency, a lesser likelihood to produce euphoria, and displaying a lower "abuse potential" (Association 1997). It is manufactured by Eli Lilly Company as Cesamet® and is available in the UK, Australia, Canada, and some European nations (Grotenhermen 2001), where it is primarily utilized as an anti-nausea agent in chemotherapy. Occasional reports have claimed benefit on spasticity in multiple sclerosis and dyskinesias. Lethal reactions have occurred in chronic canine usage (Mechoulam and Feigenbaum 1987).

Analgesic effects of nabilone in neuropathic pain patients have been noted (Notcutt, Price, and Chapman 1997), but prominent adverse effects included drowsiness and dysphoria. Some patients stated a clear preference for smoked cannabis in terms of side effects and analgesic efficacy. Nabilone's cost was estimated to be 10 times higher than herbal cannabis at black market rates, and all things considered, this agent would seem to have more disadvantages in the long term.

LEVONANTRADOL

Levonantradol is another synthetic cannabinoid from Pfizer. Analgesic benefits of up to 6 hours were noted in post-operative pain patients in a prior trial (Jain et al. 1981), but without clear dose-response effects. Adverse effects are prominent with this agent, including somnolence in 50-100% and dysphoria in 30-50% (Association 1997), termed "unacceptable" by that authority.

AJULEMIC ACID (CT3)

Ajulemic acid is a synthetic cannabinoid derived from the more stable THC-11-oic acid that does not bind to CB_1 receptors and lacks psychoactive effects. It is currently in commercial development. It has shown strong analgesic and anti-inflammatory properties in animal models of arthritis without COX-1 inhibition side effects such as ulcer production, and is advanced clinical trials (Burstein 2001, 2000). It shares anti-neoplastic effects with THC on a variety of cell lines (Recht et al. 2001), but is half as potent in this regard, although longer acting. Ajulemic acid has recently been demonstrated to bind to the peroxisome proliferator-activated receptor gamma, part of the nuclear receptor superfamily involved in inflammatory processes (Liu et al. 2003), and also to suppress human monocyte interleukin-1beta production *in vitro* (Zurier et al. 2003). Ajulemic acid portends to be a valuable addition to the pantheon of cannabinoid pharmaceuticals employed for analgesic and anti-inflammatory properties.

DEXANABINOL (HU-211)

Dexanabinol is a synthetic cannabinoid agent developed at Hebrew University from Δ^8-THC, but it is a non-psychoactive enantiomer of the fabulously potent HU-210 (Pop 2000). It has demonstrated numerous interesting properties including antioxidant and anti-inflammatory effects, as well as suppression of THF-alpha (tumor necrosis factor) production. Additionally, it reduced brain damage associated with soman (Sarin)-induced seizures in rats (Filbert et al. 1999), caused reduction of experimental autoimmune encephalomyelitis responses (Achiron et al. 2000) suggesting application in multiple sclerosis, and reduced damage in experimental focal ischemia (Lavie et al. 2001). Human trials have demonstrated mixed results. In one such Phase II study of 67 closed head injury patients, dexanabinol reduced intracranial pressure and perfusion significantly with a good adverse effect profile (Knoller et al.

2002), with some degree of improvement in clinical outcome scales after 3 and 6 months.

Dexanabinol is currently in Phase III clinical trials, and further analysis will demonstrate its relative place in the cannabinoid pharmacopoeia. As currently formulated, parenteral injection of dexanabinol is required, and it may not possess the multi-modality efficacy of Cannabis Based Medicine Extracts.

HU-308

Another agent emerging from the research of Raphael Mechoulam's laboratories in Israel is HU-308, a synthetic and specific CB_2 agonist lacking cannabinoid behavioral effects in laboratory animals (Hanus et al. 1999). Observed activities of this agent include inhibition of forskolin-stimulated cyclic AMP production, blood pressure reduction, inhibition of defecation, and production of peripheral analgesia with anti-inflammatory effects. Further testing may demonstrate an important therapeutic role for this agent.

SR141716 (RIMONABANT®)

Heretofore, our discussion has centered on cannabinoid agonists or analogues. However, given the profile of cannabinoid stimulation with its decremental effects on short-term memory acquisition and stimulation of hunger, it was expected that efforts would be mounted to clinically harness antagonistic cannabinoid effects. SR141716, dubbed Rimonabant®, is a potent CB_1-antagonist or inverse agonist used extensively in laboratory studies. It has demonstrated anti-obesity effects in mice (Ravinet Trillou et al. 2003), and is currently in human clinical trials. Preliminary results (Le Fur et al. 2001) demonstrate reduction of hunger and food intake in obese male subjects in the short term, and weight reduction in the long term, with a reportedly benign adverse effect profile. Certainly, caveats are necessary, and one might expect the emergence of depression and hyperalgesic states in patients taking this agent, such as migraine and fibromyalgia. Additionally, hypervigilance will be necessary in administering such a drug to women of child-bearing age, as SR141716 has profound effects on neonatal feeding and growth (Fride 2002).

NEW INDICATIONS FOR CANNABINOID
PHARMACEUTICALS

Emerging concepts have demonstrated the key role that endocannabinoids play in regulation of pain (Pertwee 2001), hormonal regulation
and fertility (Bari et al. 2002), hunger (Fride 2002) and gastrointestinal
function (Pertwee 2001), and even regulation of memory (Hampson and
Deadwyler 2000), and proper extinction of aversive events (Marsicano
et al. 2002).

Some of these concepts have recently been reviewed (Baker et al.
2003). In particular, the authors distinguish that cannabis and endocannabinoids may demonstrate an impairment threshold if too elevated, a
range of normal function below which a deficit threshold is breached.
This seems to be a simple and universal concept: for every neurotransmitter or neuromodulatory agent, there may be too much or too little,
with corresponding clinical pathophysiological sequelae. With respect
to endocannabinoids, this concept has been insufficiently explored. Previously, this author has postulated the likelihood of clinical endogenous
cannabinoid deficiency diseases (CECDD) (Russo 2001, 2001), including migraine, fibromyalgia, idiopathic bowel syndrome (IBS, "spastic
colon") and possible even psychiatric conditions, such as obsessive-
compulsive disorder. In light of newer information, one may posit the
addition of many other disease conditions that are seemingly unrespon-
sive to pharmacotherapy with other agents that do not influence the
endocannabinoid system: causalgia and allodynia as in brachial plexus
neuropathy and phantom limb pain, post-traumatic stress disorder (PTSD),
bipolar disease (Grinspoon and Bakalar 1998), dysmenorrhea (Russo
2002), hyperemesis gravidarum (Russo 2002; Curry 2002), unexplained
fetal wastage, glaucoma (Jarvinen, Pate, and Laine 2002), and many
others.

In the area of pain, it may be the case that we need to renew a therapeutic maneuver of the 19th century (reviewed in (Russo 2002), and
supported in (Cichewicz and Welch 2002)) by combining cannabinoids
and opioids, particularly post-operatively or in cases of major trauma,
thereby producing analgesic synergy, reducing dosages, and adverse effect profiles with respect to opiate-induced nausea, constipation and
dysphoria.

Recently, a new indication for cannabinoid manipulation has been
claimed, that of improved night vision. Based on simultaneous ethno-
botanical claims of fisherman that cannabis stimulated their ability to
see in the dark (West 1991; Merzouki and Molero Mesa 1999) in Jamaica

and Morocco, respectively, a two-pronged pilot study was launched (Russo et al. 2003). In a double-blind controlled dosage escalation study with THC as Marinol®, improvement in scotopic sensitivity was noted in one subject, while in a subsequent field study with smoked *kif* (*Cannabis sativa/Nicotiana rustica* mixture) in three subjects, improvement in both dark adaptation and scotopic sensitivity thresholds were noted with the SST-1 Scotopic Sensitivity Tester (Peters, Locke, and Birch 2000). Given the relative paucity of CB_1 receptors in the striate cortex (Glass, Dragunow, and Faull 1997), and their particular density in rod spherules (Straiker et al. 1999), this phenomenon seems to be of retinal, rather than cortical origin. This is further supported by anecdotal claims that cannabis improves vision in retinitis pigmentosa (Arnold 1998). Based on these findings, more formal studies of RP with fully objective measures such as electroretinography seem warranted. Given the neuroprotective and antioxidant effects of cannabis and cannabinoids, extension of therapy to senile macular degeneration appears most promising.

CANNABINOIDS AND NEUROPROTECTION

In light of recent demonstration of the ability of THC and CBD to prevent cell death from glutamate toxicity (Hampson et al. 1998), a whole host of new therapeutic applications gain more than theoretical support beyond the current studies of stroke and closed head injury discussed in relation to dexanabinol. Therapeutic claims for cannabis in amyotrophic lateral sclerosis (ALS) have been advanced in a single case study (Carter and Rosen 2001), and it may prove to be that neurodegeneration may be diminished or arrested in this disorder, Huntington disease (Glass 2001), Parkinson disease (Sieradzan et al. 2001), Alzheimer disease (Volicer et al. 1997), and others. Neuroprotection is a valuable effect, as well, in treatment of seizure disorders (Cunha et al. 1980; Carlini and Cunha 1981; Wallace, Martin, and DeLorenzo 2002). The role of cannabis therapeutics in HIV encephalopathy and slow virus (prion) diseases (Bovine Spongiform Encephalopathy (BSE) or "mad cow disease," Creutzfeldt-Jakob disease, etc.) deserves exploration based on these preliminary findings.

Emerging concepts in psychiatry support that depression is not merely attributable to deficiencies of serotonin, norepinephrine or dopamine (Delgado and Moreno 1999), but rather, may represent a disorder of neuroplasticity suggesting the desirability to employ neuroprotective agents. An extensive history of such use over the last 4000 years (Russo 2001), coupled with this new information, lends credence to the hypothesis. With their unique pharmacological profiles, CBMEs deserve an effort in clinical trials.

SPASMODIC DISORDERS

The current information supporting muscle relaxant benefits of cannabis and cannabinoids in MS and spinal cord injury is extremely compelling. Mining the data of the past (O'Shaughnessy 1838-1840; Christison 1851; Reynolds 1868, 1890), one may wonder anew about the role of cannabinoid therapeutics in disorders such as tetanus, hiccup (Gilson and Busalacchi 1998), stiff man syndrome, the various periodic paralyses, and dystonic disorders such as torticollis, dystonia musculorum deformans, stuttering, and writer's cramp.

FORBIDDEN TERRITORIES

Obstetrics and Gynecology

This topic has been recently reviewed at length (Russo 2002; Russo, Dreher, and Mathre 2003). Cannabis has been employed for millennia for a variety of related ills. Drugs are rightly eschewed when possible in pregnancy, but cases arise frequently wherein such treatment is necessary, even to save the life of mother and child. Close scrutiny of the literature supports the relative safety of cannabis in such applications, and particularly in episodic use, it is highly likely that the cost-benefit ratio in serious disorders is quite acceptable. Controlled studies of dysmenorrhea, hyperemesis gravidarum and other disorders with cannabis-extracts and medicines should be advanced.

Cannabinoid Medicines in Pediatrics

It is clear that cannabis and cannabinoids hold promise in for many intractable and desperate pediatric conditions, although this concept may be anathema to some.

Although it is frequently the butt of jokes, no one who has not been the parent of an affected infant can truly conceive of the stress and disturbance engendered by infantile colic. A developmental disorder appearing most often between two weeks and three months of life, this poorly understood syndrome produces nightly bouts of inconsolable crying and apparent abdominal cramping pain. Myriad remedies aimed at every imaginable neurotransmitter system of brain and gut tend to fail to stem its ravages. Perhaps infantile colic is another developmental clinical endogenous cannabinoid deficiency disorder. With its anti-spasmodic, analgesic, anti-anxiety and soporific attributes, a THC:CBD cannabis extract holds promise where other agents have disappointed, and if so, countless new parents may be thankful.

Another possible pediatric indication for cannabis-based medicines is cystic fibrosis. In a recent study (Fride 2002), an extremely compelling and well-conceived rationale for cannabis treatment was outlined that could vastly improve the clinical condition and well-being of affected children. Similar benefits might accrue to other serious failure-to-thrive states.

Cannabis medicines have already demonstrated remarkable success in allaying nausea and vomiting in children undergoing cancer chemotherapy (Abrahamov and Mechoulam 1995). Unfortunately, this study has been largely ignored, rather than being duplicated and extended. Any possible moral objection to such treatment holds no weight when the alternative is severe suffering and even death of a child. The recent report of cannabidiol (CBD) inhibition of glioma cell growth by promotion of apoptosis independent of cannabinoid and vanilloid receptor activity (Vaccani, Massi, and Parolaro 2003), should convince all but the most hardened detractors.

A less lethal, but yet still compelling potential indication is childhood asthma. The advent of new delivery devices for cannabis medicines discussed in this volume, combining bronchodilation, with modulation of leukotrienes and other mediators of inflammation offer unique benefits to this disorder.

Finally, the area of child psychiatry deserves additional consideration. A recent book, *Jeffrey's Journey: A Determined Mother's Battle for Medical Marijuana for Her Son* (Jeffries and Jeffries 2003), documents the case study of a young man who failed every conceivable psychopharmacological agent to control his anger and other psychopathology. Only oral cannabis worked, preventing his imminent institutionalization, and allowing a return to a semblance of normal life.

This author, in his practice of child and adult neurology, has heard dozens of unsolicited testimonials to the benefits of cannabis in attention-deficit hyperactivity disorder (ADHD), supporting available anecdotal accounts (Grinspoon and Bakalar 1997). Although the idea of using cannabis-based medicines for this indication may seem surprising to most experts, controlled trials of cannabis medicines for children with ADHD seem clearly indicated, particularly in view of the controversies and side effects of existing psychotropic medications. Extension of the concept to other difficult disorders of obscure pathophysiology such as autistic spectrum and Asperger disorders may be warranted. If and when cannabis establishes its efficacy in pediatric diseases, it shall have achieved a fair measure of redemption from the derision it has elicited during the past century.

REFERENCES

Abrahamov, A., and R. Mechoulam. 1995. An efficient new cannabinoid antiemetic in pediatric oncology. *Life Sci* 56 (23-24):2097-102.

Achiron, A., S. Miron, V. Lavie, R. Margalit, and A. Biegon. 2000. Dexanabinol (HU-211) effect on experimental autoimmune encephalomyelitis: implications for the treatment of acute relapses of multiple sclerosis. *J Neuroimmunol* 102 (1):26-31.

Arnold, S. 1998. Seeing is believing. *Nurs Stand* 12 (22):17.

Association, British Medical. 1997. *Therapeutic uses of cannabis*. Amsterdam: Harwood Academic Publishers.

Baker, D., G. Pryce, G. Giovannoni, and A. J. Thompson. 2003. The therapeutic potential of cannabis. *Lancet Neurol* 2(May):291-298.

Bari, M., N. Battista, A. Cartoni, G. D'Arcangelo, and M. Maccarone. 2002. Endocannabinoid degradation and human fertility. *J Cannabis Therapeutics* 2 (3-4):37-49.

Burstein, S. H. 2000. Ajulemic Acid (CT3): A Potent Analog of the Acid Metabolites of THC. *Curr Pharm Res* 6 (13):1339-1345.

_____2001. The therapeutic potential of ajulemic acid (CT3). In *Cannabis and cannabinoids: Pharmacology, toxicology and therapeutic potential*, edited by F. Grotenhermen and E. Russo. Binghamton, NY: Haworth Press.

Carlini, E. A., and J. M. Cunha. 1981. Hypnotic and antiepileptic effects of cannabidiol. *J Clin Pharmacol* 21 (8-9 Suppl):417S-427S.

Carter, G. T., and B. S. Rosen. 2001. Marijuana in the management of amyotrophic lateral sclerosis. *Am J Hosp Palliat Care* 18 (4):264-70.

Christison, A. 1851. On the natural history, action, and uses of Indian hemp. *Monthly J Medical Science of Edinburgh, Scotland* 13:26-45, 117-121.

Cichewicz, D. L., and S. P. Welch. 2002. The effects of oral administration of delta-9-THC on morphine tolerance and physical dependence. Paper read at Symposium on the Cannabinoids, July 13, at Asilomar Conference Center, Pacific Grove, CA.

Cunha, J. M., E. A. Carlini, A. E. Pereira, O. L. Ramos, C. Pimentel, R. Gagliardi, W. L. Sanvito, N. Lander, and R. Mechoulam. 1980. Chronic administration of cannabidiol to healthy volunteers and epileptic patients. *Pharmacol* 21(3):175-85.

Curry, W.-N. L. 2002. Hyperemesis gravidarum and clinical cannabis: To eat or not to eat? *J Cannabis Therapeutics* 2 (3-4):63-83.

172   CANNABIS: FROM PARIAH TO PRESCRIPTION

Delgado, P., and F. Moreno. 1999. Antidepressants and the brain. *Int Clin Psychopharmacol* 14 Suppl 1:S9-16.

Filbert, M. G., J. S. Forster, C. D. Smith, and G. P. Ballough. 1999. Neuroprotective effects of HU-211 on brain damage resulting from soman-induced seizures. *Ann N Y Acad Sci* 890:505-14.

Fride, E. 2002. Cannabinoids and cystic fibrosis: A novel approach. *J Cannabis Therapeutics* 2 (1):59-71.

_____2002. Cannabinoids and feeding: The role of the endogenous cannabinoid system as a trigger for newborn suckling. *J Cannabis Therapeutics* 2 (3-4):51-62.

Gilson, I., and M. Busalacchi. 1998. Marijuana for intractable hiccups. *Lancet* 351 (9098):267.

Glass, M. 2001. The role of cannabinoids in neurodegenerative diseases. *Prog Neuropsychopharmacol Biol Psychiatry* 25 (4):743-65.

Glass, M., M. Dragunow, and R. L. Faull. 1997. Cannabinoid receptors in the human brain: A detailed anatomical and quantitative autoradiographic study in the fetal, neonatal and adult human brain. *Neurosci* 77 (2):299-318.

Grinspoon, L., and J. B. Bakalar. 1998. The use of cannabis as a mood stabilizer in bipolar disorder: Anecdotal evidence and the need for clinical research. *J Psychoactive Drugs* 30 (2):171-7.

Grinspoon, L., and J. B. Bakalar. 1997. *Marihuana, the forbidden medicine*. Rev. and exp. ed. New Haven: Yale University Press.

Grotenhermen, F. 2001. Definitions and explanations. In *Cannabis and cannabinoids: Pharmacology, toxicology and therapeutic potential*, edited by F. Grotenhermen and E. Russo. Binghamton, NY: Haworth Press.

Hampson, A. J., M. Grimaldi, J. Axelrod, and D. Wink. 1998. Cannabidiol and (-)Delta9-tetrahydrocannabinol are neuroprotective antioxidants. *Proc Natl Acad Sci USA* 95 (14):8268-73.

Hampson, R. E., and S. A. Deadwyler. 2000. Cannabinoids reveal the necessity of hippocampal neural encoding for short-term memory in rats. *J Neurosci* 20 (23): 8932-42.

Hanus, L., A. Breuer, S. Tchilibon, S. Shiloah, D. Goldenberg, M. Horowitz, R. G. Pertwee, R. A. Ross, R. Mechoulam, and E. Fride. 1999. HU-308: A specific agonist for CB(2), a peripheral cannabinoid receptor. *Proc Natl Acad Sci USA* 96 (25):14228-33.

Jain, A. K., J. R. Ryan, F. G. McMahon, and G. Smith. 1981. Evaluation of intramuscular levonantradol and placebo in acute postoperative pain. *J Clin Pharmacol* 21 (8-9 Suppl):320S-326S.

Jarvinen, T., D. Pate, and K. Laine. 2002. Cannabinoids in the treatment of glaucoma. *Pharmacol Ther* 95 (2):203.

Jeffries, D., and L. Jeffries. 2003. *Jeffrey's journey: A determined mother's battle for medical marijuana for her son*. Rocklin, CA: LP Chronicles.

Knoller, N., L. Levi, I. Shoshan, E. Reichenthal, N. Razon, Z. H. Rappaport, and A. Biegon. 2002. Dexanabinol (HU-211) in the treatment of severe closed head injury: A randomized, placebo-controlled, phase II clinical trial. *Crit Care Med* 30 (3):548-54.

Lavie, G., A. Teichner, E. Shohami, H. Ovadia, and R. R. Leker. 2001. Long term cerebroprotective effects of dexanabinol in a model of focal cerebral ischemia. *Brain Res* 901 (1-2):195-201.

Le Fur, G., M. Arnone, M. Rinaldi-Carmona, F. Barth, and H. Heshmati. 2001. SR141716, a selective antagonist of CB1 receptors and obesity. Paper read at Symposium on the Cannabinoids, June 29, at El Escorial, Spain.

Liu, J., H. Li, S. H. Burstein, R. B. Zurier, and J. D. Chen. 2003. Activation and binding of peroxisome proliferator-activated receptor gamma by synthetic cannabinoid ajulemic acid. *Mol Pharmacol* 63 (5):983-92.

Marsicano, G., C. T. Wotjak, S. C. Azad, T. Bisogno, G. Rammes, M. G. Cascio, H. Hermann, J. Tang, C. Hofmann, W. Zieglgansberger, V. Di Marzo, and B. Lutz. 2002. The endogenous cannabinoid system controls extinction of aversive memories. *Nature* 418 (6897):530-4.

Mechoulam, R., and J. J. Feigenbaum. 1987. Toward cannabinoid drugs. In *Progress in medicinal chemistry*, edited by G. Ellis and G. West. Amsterdam: Elsevier Science.

Merzouki, A., and J. Molero Mesa. 1999. La [sic] chanvre *(Cannabis sativa* L.) dans la pharmacopée traditionelle du Rif (nord du Maroc). *Ars Pharmaceutica* 40 (4):233-240.

Notcutt, William, Mario Price, and Glen Chapman. 1997. Clinical experience with nabilone for chronic pain. *Pharmaceut Sci* 3:551-555.

O'Shaughnessy, W. B. 1838-1840. On the preparations of the Indian hemp, or gunjah *(Cannabis indica)*; Their effects on the animal system in health, and their utility in the treatment of tetanus and other convulsive diseases. *Transactions of the Medical and Physical Society of Bengal*:71-102, 421-461.

Pertwee, R. G. 2001. Cannabinoid receptors and pain. *Prog Neurobiol* 63 (5):569-611.

_____2001. Cannabinoids and the gastrointestinal tract. *Gut* 48 (6):859-67.

Peters, A. Y., K. G. Locke, and D. G. Birch. 2000. Comparison of the Goldmann-Weekers dark adaptometer and LKC Technologies Scotopic Sensitivity tester-1. *Doc Ophthalmol* 101 (1):1-9.

Pop, E. 2000. Dexanabinol Pharmos. *Curr Opin Investig Drugs* 1 (4):494-503.

Ravinet Trillou, C., M. Arnone, C. Delgorge, N. Gonalons, P. Keane, J. P. Maffrand, and P. Soubrie. 2003. Anti-obesity effect of SR141716, a CB1 receptor antagonist, in diet-induced obese mice. *Am J Physiol Regul Integr Comp Physiol* 284 (2): R345-53.

Recht, L. D., R. Salmonsen, R. Rosetti, T. Jang, G. Pipia, T. Kubiatowski, P. Karim, A. H. Ross, R. Zurier, N. S. Litofsky, and S. Burstein. 2001. Antitumor effects of ajulemic acid (CT3), a synthetic non-psychoactive cannabinoid. *Biochem Pharmacol* 62 (6):755-63.

Reynolds, J. R. 1868. On some of the therapeutical uses of Indian hemp. *Arch Med* 2:154-160.

_____1890. Therapeutical uses and toxic effects of *Cannabis indica*. *Lancet* 1: 637-638.

Russo, E. 2002. Cannabis treatments in obstetrics and gynecology: A historical review. *J Cannabis Therapeutics* 2 (3-4):5-35.

Russo, E. B. 2002. Role of cannabis and cannabinoids in pain management. In *Pain management: A practical guide for clinicians*, edited by R. S. Weiner. Boca Raton, FL: CRC Press.

Russo, E. B., M. Dreher, and M. L. Mathre. 2003. *Women and cannabis: Medicine, science, and sociology*. Binghamton, NY: Haworth Press.

Russo, E. B., A. Merzouki, J. Molero Mesa, and K. A. Frey. 2003. Cannabis improves night vision: A pilot study of dark adaptometry and scotopic sensitivity in kif smokers of the Rif Mountains of Northern Morocco. *J Ethnopharmacol* (Submitted).

Russo, E. B. 2001. *Handbook of psychotropic herbs: A scientific analysis of herbal remedies for psychiatric conditions.* Binghamton, NY: Haworth Press.

_____2001. Hemp for headache: An in-depth historical and scientific review of cannabis in migraine treatment. *J Cannabis Therapeutics* 1 (2):21-92.

Sieradzan, K. A., S. H. Fox, M. Hill, J. P. Dick, A. R. Crossman, and J. M. Brotchie. 2001. Cannabinoids reduce levodopa-induced dyskinesia in Parkinson's disease: A pilot study. *Neurol* 57(11):2108-11.

Straiker, A., N. Stella, D. Piomelli, K. Mackie, H. J. Karten, and G. Maguire. 1999. Cannabinoid CB1 receptors and ligands in vertebrate retina: Localization and function of an endogenous signaling system. *Proc Natl Acad Sci USA* 96 (25):14565-70.

Vaccani, A., P. Massi, and D. Parolaro. 2003. Inhibition of human glioma cell growth by the nonpsychoactive cannabidiol. Paper read at First European Workshop on Cannabinoid Research, April 4-5, at Madrid.

Volicer, L., M. Stelly, J. Morris, J. McLaughlin, and B. J. Volicer. 1997. Effects of dronabinol on anorexia and disturbed behavior in patients with Alzheimer's disease. *Int J Geriatr Psychiatry* 12 (9):913-9.

Wallace, M. J., B. R. Martin, and R. J. DeLorenzo. 2002. Evidence for a physiological role of endocannabinoids in the modulation of seizure threshold and severity. *Eur J Pharmacol* 452 (3):295-301.

West, M. E. 1991. Cannabis and night vision. *Nature* 351:703-704.

Zurier, R. B., R. G. Rossetti, S. H. Burstein, and B. Bidinger. 2003. Suppression of human monocyte interleukin-1beta production by ajulemic acid, a nonpsychoactive cannabinoid. *Biochem Pharmacol* 65 (4):649-55.

Index

ABS Laboratories Ltd., 84
ADHD. *See* Attention-deficit
    hyperactivity disorder
    (ADHD)
ADS. *See* Advanced Delivery System
    (ADS)
Adult Reading Test, 47
Advanced Delivery System (ADS), 2
    diagram of, 12,15f
    sublingual spray in, 11,14f
Advanced Dispensing System (ADS),
    11,14f
Aerosol, pressurised, in single centre,
    placebo-controlled four
    period, crossover, tolerability
    study of three formulations of
    CBMEs, 58-59,59f,63-65,
    64t,65t,70
Ajulemic acid, future of, 165
Alpha-pinene, 11
ALS. *See* Amyotrophic lateral
    sclerosis (ALS)
Alzheimer disease, 168
American Institute of Medicine, 14
Amyotrophic lateral sclerosis (ALS),
    168
Atharva Veda, 19
Attention-deficit hyperactivity disorder
    (ADHD), 170
AUC_{0-t}, defined, 52t
Audit(s)
    clinical quality, 133
    quality assurance, in single centre,
    placebo-controlled four
    period, crossover, tolerability
    study of three formulations of
    CBMEs, 51

BDS. *See* Botanical Drug Substance
    (BDS)
Beverage(s)
    in open label, four-way crossover
    study of pharmacokinetic
    profiles of CBMEs, 92-93
    in phase I, double-blind three-way
    crossover study of
    pharmacokinetic profile of
    CBMEs, 131,132t
    in single centre, placebo-controlled
    four period, crossover,
    tolerability study of three
    formulations of CBMEs,
    49-50,50t
Bipolar disease, 167
Blood pressure
    in phase I, double-blind three-way
    crossover study of
    pharmacokinetic profile of
    CBMEs, 134
    in single centre, placebo-controlled
    four period, crossover,
    tolerability study of three
    formulations of CBMEs,
    53,69
Blood pressure measurement, in single
    centre, placebo-controlled
    four period, crossover,
    tolerability study of three
    formulations of CBMEs, 48
Blood sampling for plasma concentration
    analysis, in single centre,
    placebo-controlled four
    period, crossover, tolerability
    study of three formulations of
    CBMEs, 48

http://www.haworthpress.com/store/product.asp?sku=J175
© 2003 by The Haworth Press, Inc. All rights reserved.

SPECIAL 25%-OFF DISCOUNT!

Order a copy of this book with this form or online at:
http://www.haworthpress.com/store/product.asp?sku=5124
Use Sale Code BOF25 in the online bookshop to receive 25% off!

Cannabis

From Pariah to Prescription

____ in softbound at $14.96 (regularly $19.95) (ISBN: 0-7890-2399-7)
____ in hardbound at $29.96 (regularly $39.95) (ISBN: 0-7890-2398-9)

COST OF BOOKS ____	❏ **BILL ME LATER:** ($5 service charge will be added)
Outside USA/ Canada/ Mexico: Add 20%. ____	Bill-me option is good on US/Canada/ Mexico orders only; not good to jobbers, wholesalers, or subscription agencies.
POSTAGE & HANDLING ____	
US: $4.00 for first book & $1.50 for each additional book	❏ **Signature** ____
Outside US: $5.00 for first book & $2.00 for each additional book.	❏ **Payment Enclosed: $** ____
SUBTOTAL ____	❏ **PLEASE CHARGE TO MY CREDIT CARD:**
In Canada: add 7% GST. ____	❏ Visa ❏ MasterCard ❏ AmEx ❏ Discover ❏ Diner's Club ❏ Eurocard ❏ JCB
STATE TAX ____	**Account #** ____
CA, IN, MIN, NY, OH, & SD residents please add appropriate local sales tax.	
FINAL TOTAL ____	**Exp Date** ____
If paying in Canadian funds, convert using the current exchange rate, UNESCO coupons welcome.	**Signature** ____
	(Prices in US dollars and subject to change without notice.)

PLEASE PRINT ALL INFORMATION OR ATTACH YOUR BUSINESS CARD
Name
Address
City   State/Province   Zip/Postal Code
Country
Tel   Fax
E-Mail

May we use your e-mail address for confirmations and other types of information? ❏Yes❏ No
We appreciate receiving your e-mail address. Haworth would like to e-mail special discount offers to you, as a preferred customer. **We will never share, rent, or exchange your e-mail address.** We regard such actions as an invasion of your privacy.

Order From Your Local Bookstore or Directly From
The Haworth Press, Inc.
10 Alice Street, Binghamton, New York 13904-1580 • USA
Call Our toll-free number (1-800-429-6784) / Outside US/Canada: (607) 722-5857
Fax: 1-800-895-0582 / Outside US/Canada: (607) 771-0012
E-Mail your order to us: Orders@haworthpress.com

Please Photocopy this form for your personal use.
www.HaworthPress.com

BOF03